Magnifications

David Scharf

Magnifications

Photography with the
Scanning Electron
Microscope

Schocken Books • New York

First published by SCHOCKEN BOOKS 1977
First SCHOCKEN PAPERBACK edition 1978

Copyright © 1977 by David Scharf

Library of Congress Cataloging in Publication Data
Scharf, David, 1942–
  Magnifications.

  1. Photomicrography.   2. Scanning electron
microscope.   I. Title.
QH251.S32   778.3′1   77-75283

Manufactured in the United States of America

CONTENTS

I wish to thank the following people for their kind help: Dr. Charles L. Hogue, Dr. William A. Emboden Jr., and Dr. Roy Snelling for their assistance in identifying certain specimens; and my brother Barry for his assistance with the illustrations.

This book is affectionately dedicated to my wife, Patty, and my daughter, Edith Anne.

FOREWORD

With the planets, stars, and galaxies in one direction and the small life forms, molecules, atoms, and subatomic particles in the other, we are, for all earthly purposes, somewhere in the middle of an infinity in terms of physical dimensions. Our size and the limits of our physical senses tend to confine our experience to a certain niche of the universe. Our eyes, in particular, respond to only a very limited portion of the known electromagnetic spectrum, which we, of course, call "light". On one end of this spectrum we find ultraviolet light, x-rays, gamma rays, and cosmic rays—the frequency or vibration of these becoming increasingly higher and their wavelength increasingly shorter. At the other end of the spectrum, infrared, microwaves, radio waves, and waves created by electric power generation have increasingly lower frequencies and longer wavelengths.

The point is that the wavelength of visible light sets a limit upon the detail or resolution that can be perceived by our eyes. To the normal, unaided eye this limit is imperceptible. But when we begin looking through optical microscopes, the useful limit or ordinary light magnification soon becomes apparent.

In this technological era, Man's impulse to expand his consciousness is evident in the myriad, exotic instruments he has created to take him beyond the normal limits of his senses. Until comparatively recently, these instruments were fairly simple and one could readily comprehend their function. But in the last seventy-five years or so, a new generation of instruments has come into being, founded in logic and analogy and designed in abstraction by mathematics and the theories of physical law—the science of physics having just been revolutionized by quantum mechanics and relativity. From simple to highly complex: the concept of systems integration—taking several simple devices, organizing them into a system, then integrating several systems to utilize their net effect—has been responsible for the modern, highly sophisticated generation of search and research instruments. As the function of our tools has moved from direct to indirect sensing, so does the experience move from the direct to the indirect. Nevertheless, these tools have taken us into heretofore unknown places.

The scanning electron microscope is one such tool, which has enabled us to take a closer look at the smaller regions of the cosmos. Here a strange, silent beauty is encountered; we see in a way that the eyes by themselves cannot, the forms, shapes, and patterns of nature—some of which have never been dreamed of before, while others seem to ring with vague familiarity.

The primary purpose of this book is to provide a visual experience of this new imagery. It is not intended to provide a representative cross-section of plant and animal life, nor is it a scientific documentation of all aspects of the scanning electron microscope. Photographs of biological subjects were made using my own techniques, which enable me to record many types of living things in their natural state—the subjects were alive and untreated—whereas previously, most S.E.M. photography was performed with organic subjects which were dead, dried, and coated with a gold alloy. I did coat some *inanimate* objects in the customary manner as indicated in the captions.

"Lighting" for the photographs is provided by *electron* illumination and not light, hence, there is no color involved.

A HISTORY OF THE MICROSCOPE

The magnifying lens has been around since ancient times; records date it back to 1,000 B.C. or more and no one knows just how long ago a drop of water was used to magnify. The Assyrians had a double convex lens around 700 B.C. The Greeks and Romans also wrote of the lens.

The compound microscope is a device which utilizes two or more lenses. Its story begins about 1590 in Holland where it was invented by a Dutch spectacle maker named Zacharias Janssen. It was said to have been an inch in diameter and six feet long. This discovery started a parade of inventors and improvements in the microscope. In 1610 Gallileo improved the microscope by providing a screw thread as a means to focus. Many others also made microscopes in this early period including Anton Von Leeuwenhoek, who made over 100 microscopes with simple lenses, some of which were ⅛″ or so in diameter with a very strong curvature and thus high magnification. With these instruments he discovered bacteria and the existence of corpuscles in the blood.

In the last quarter of the 1600s, Robert Hooke made a three lens compound microscope which was easy to use. The Royal Microscope Society of England published Hooke's "Micrographia", in which he describes tissues, blood vessels, textiles, crystals, and insect structures.

The lens was further improved in the mid-eighteenth century by John Dollond who devised the "achromatic" or color corrected lens. It was not used in a microscope, however, until the early nineteenth century. Then in 1873 Ernst Abbe wrote "The Theory of Microscope Image Formation", which resulted in bringing the design of the compound microscope to its peak. Progress in optical microscopy since that time has been mainly in the nature of refinements, mass production, and techniques of usage.

By the early 1900s, another possibility was

available. It was seen that electrons could be guided in curved paths and thus used to magnify images. By 1931, the first electron microscope was built in Germany; in 1939, the Germans had a commercial model available. This was the transmission type of electron microscope.

The principle for a scanning electron microscope was first described by H. Stintzing in 1927. In 1935 it was demonstrated by M. Knoll. Thirty years later, in 1965, after much development by Professor Oatley and his associates at Cambridge, England, the first commercial scanning electron microscope became available. By 1975, it had been greatly refined, with instruments of high resolution and high information-density recording ability available.

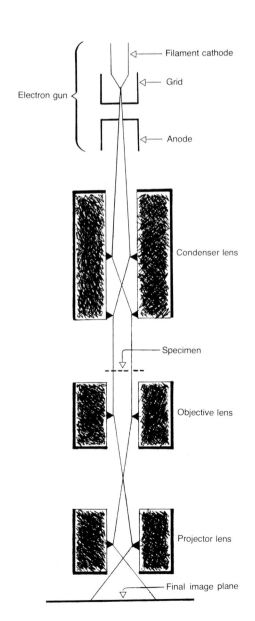

Direct Electron Microscope (Transmission E.M.–T.E.M.)

Labels on diagram:
- Filament cathode
- Grid
- Electron gun
- Anode
- Condenser lens
- Specimen
- Objective lens
- Projector lens
- Final image plane

THE S.E.M.

In discussing the relative capabilities of different kinds of microscopes, it is necessary to introduce the Angstrom (Å) as a basic unit of measurement. An Angstrom unit is equal to one ten-billionth (.0000000001) of a meter or four billionths (.000000004) of an inch. It is used to measure the *resolution*—that is, the ability to perceive and separate detail—of microscopes.

We begin with a comparison of microscope types:

Optical Microscopes are generally limited to about a 5,000Å resolution or about 2,000X; with special equipment and techniques, using ultraviolet light, this may be extended to around 2,000Å or 5,000X. At moderate and high magnification, the optical microscope has extremely poor depth-of-focus, making it quite difficult to view any complex three-dimensional surface features.

The Direct Electron Microscope, usually referred to as a **Transmission Electron Microscope** or T.E.M., has the greatest available resolution; 10Å being readily obtained from commercial instruments and 2Å or better possible with the best equipment and techniques. This instrument

A comparatively new model scanning electron microscope

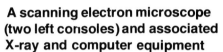

A scanning electron microscope (two left consoles) and associated X-ray and computer equipment

accepts only thinly sliced specimens or replicas and has only one main mode of imagery. It works by projecting an electron beam through a thin section of specimen, whereupon the resultant beam is focused on a fluorescent screen to an image which is an "electron shadow" of the specimen. The depth of focus is moderate.

The Scanning Electron Microscope (S.E.M.) in its normal mode of imagery—that of secondary electrons—is used to examine the *exterior* surface of objects as contrasted with the transmission electron microscope's probing of inner structure. A 100Å resolution is routinely attained in most high quality S.E.M.s. In the best

commercial S.E.M.s, having field emission electron sources, 25Å is attainable. Very important is the ability to use this instrument at low magnification—from 5X or so. At the lower magnification range of optical instruments, there is the advantage of having the entire specimen in focus, since the depth-of-focus is very great, and of having a much greater ultimate resolution of detail. One can readily zoom in or out with a 10,000 to 1 ratio. Also, the S.E.M. will accept specimens which are intact.

Just recently, research S.E.M.s have attained a resolution better than 2Å with magnifications of six million or so. They have been used to image individual atoms (in the transmission mode).

3

How the S.E.M. works

A Simplified Explanation. In very simple terms, the scanning electron microscope, as compared with an optical microscope, uses electrons instead of light and its lenses are magnetic rather than glass. As previously mentioned, the S.E.M. is used to image the surface features of objects. This is accomplished by scanning a finely focused beam of electrons across the surface of a specimen in a high vacuum. As the beam of electrons strikes the object's surface, it causes other electrons to be knocked out into the vacuum where they are attracted to a sensor. There the electrons are detected, converted to a data signal, processed, and then reassembled as a picture on a high resolution television monitor. At this point, the image may be viewed or photographically recorded.

In the S.E.M., only one spot on the subject's surface need be in focus at any one time. Hence, the S.E.M. is able to vary the focus point constantly to follow the contours of the surface— what is known as "dynamic focus"—and thus,

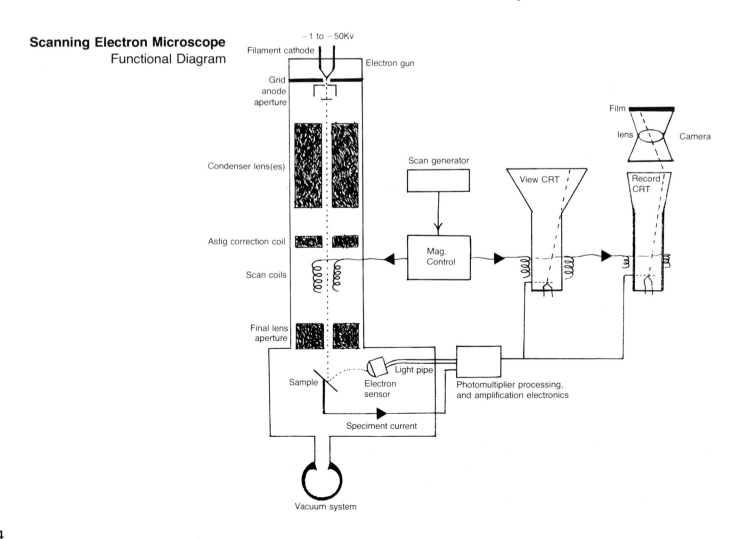

Scanning Electron Microscope
Functional Diagram

in theory, produce an overall image that is entirely in focus from point to point. As an added possibility, since the image information is carried by electronic signal after being detected, the data may be processed, enhanced, or stored in a computer, if necessary.

A More Detailed Explanation. The S.E.M's basic components are as follows:

- An electron source;
- A means of forming the electrons into a beam and focusing it to a tiny spot on the specimen;
- A means to scan the spot across the specimen;
- A means of detecting a response from the specimen;
- A means of transmitting the response from the specimen to the display system; and
- A display system whose scan is in synchronization with the primary beam.

The system, from the electron source, through the electron-optical column, to the specimen chamber, is under high vacuum—about one millionth of an atmospheric pressure. This is to allow the electrons a free path so as not to collide with air molecules; collision with air molecules would make a very noisy signal, or if present in great enough quantity, would even prevent the beam from reaching the specimen.

The Electron Source: A pointed tungsten *filament* is generally used; it is heated to a high temperature by passing a current through it, and a high, negative polarity accelerating voltage is applied (normally 1,000 to 30,000 volts). The high voltage, the heating of the filament, and the tendency of a charge to accumulate at a point, all contribute to cause electrons to be emitted into the vacuum. The action of the *grid* tends to concentrate the emission of electrons at the filament's point; the *anode* helps to extrude the electrons and direct them; and the series of holes in the *grid, anode,* and *spray aperture* helps to begin the formation of the beam. All this in combination is called the *electron gun.*

Magnetic Lenses: The electrons are directed down the *electron-optical column* through the *condenser lens* regions which form them into a beam and focus it. The next lens encountered is the *stigmator,* which corrects for any resultant

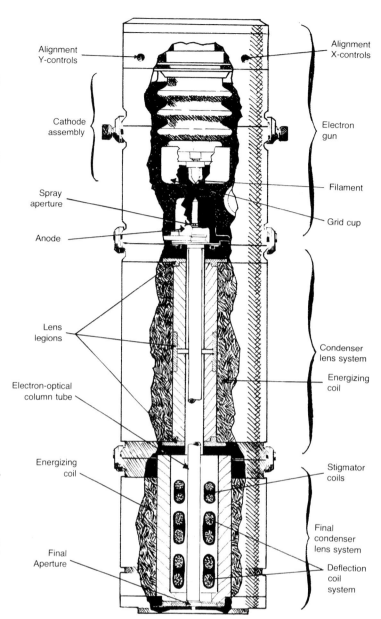

Electron optical column

5

astigmatic aberrations in the beam; then through the *scanning coil* regions which scans the beam across the specimen. In some instruments the scanning coils are first. At last, the beam goes through the *final aperture* and then through the region of the *final lens,* which provides the final focus of the beam to a tiny spot on the sample.

Vacuum Chamber (door open). Note mechanical stage on door. Detector, is visible inside of chamber (with screen).

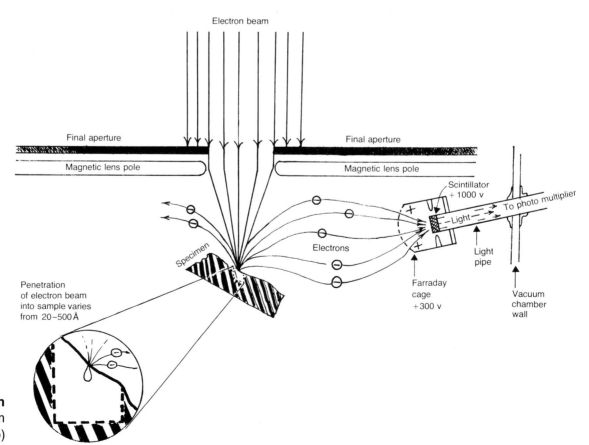

Detector illustration (print with vacuum chamber photo)

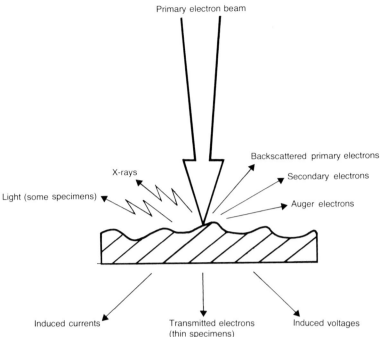

Primary electron beam

Backscattered primary electrons

X-rays

Secondary electrons

Light (some specimens)

Auger electrons

Induced currents

Transmitted electrons
(thin specimens)

Induced voltages

Detection and Amplification: The electron beam strikes the specimen and interacts with its surface, knocking "secondary" electrons out of the orbits of the atoms of specimen material. These secondary electrons along with recoiled primary electrons are emitted out into the vacuum where they are attracted to the *detector* by its opposite (+) charge as shown in the illustration. The electrons strike the *scintillator*, which is coated with a phosphor; the collisions create tiny flashes of light, which are then directed out of the vacuum chamber via a *light pipe*, to the *photomultiplier tube*, a device which converts the light back to an electronic signal and amplifies it. This signal is now further amplified, processed, and enhanced as desired, after which time it is transmitted to a *cathode ray tube* (television monitor—C.R.T.) which has its own electron beam that is in synchronization with the primary beam of the electron-optical column. Both of these beams are controlled by the *scan generator*. The transmitted data signal modulates intensity of the C.R.T.'s beam which then writes the information upon the phosphor screen as a picture. Generally there are two C.R.T.s—one for viewing and the other for high resolution photographic recording.

Since the size of the image on the view and record C.R.T.s is constant, all that is necessary to increase magnification is to decrease the area of specimen being scanned.

The (above) illustration shows the various types of excitations and emissions caused by the interaction of the electron beam with the specimen.

Each of these excitations may be used to produce imagery when sensed by an appropriate detector; they provide additional information about the nature of the material being studied: x-rays and Auger electrons indicate the presence of a particular chemical element; induced currents and voltages reveal electrical properties; electron channeling patterns (not illustrated) give information about the crystal lattice of the specimen. Additionally, emitted electrons may be collected with respect to their angle of emission or their energy; some materials, like phosphors, emit light when bombarded by electrons; and thin specimens may be imaged via transmitted electrons. In this mode of operation the instrument is referred to as a scanning transmission electron microscope (S.T.E.M.).

Besides creating imagery from the various specimen-electron interactions, one may process the detected signal to obtain other image displays: line scan, dot scan, amplitude modulated (Y-modulation), derivative image, negative image, and three-dimensional image via stereo-pairs.

S.E.M. Techniques

Biological specimens intended for observation in the S.E.M. are generally prepared by one or more of several methods. The preparation technique usually depends upon the nature of the sample—whether it is an internal or external structure and whether it is a wet or dry surface. Some of these techniques are: chemical fixation and staining, solvent solutions for clearing and washing, and chemical etching; then, dehydration by either air drying, air drying with the use of solvent baths, freeze drying, or critical-point drying; after which, a thin layer of carbon, gold, or some other conductive material is applied by vacuum evaporation to the surface to accept the electron beam. This preparation "fixes" the specimen so that it may be observed easily and almost indefinitely by standard observation procedures.

For all of their benefits, however, these preparation techniques have at least one drawback. They subject the specimens to some degree of trauma—sometimes a great degree—thus, the pictures which are obtained are not usually a very accurate representation of the natural state: the surface is likely to buckle, shrivel, or crack; as well, many an interesting find may be obliterated. When the gold is deposited, there may at times be shadows cast due to the angle of deposition and/or sample surface topography. The gold does not always penetrate all the folds and crevices of detail, thus obtaining a uniform gold deposition on a complex topography is extremely difficult. Depositing a very heavy layer will partially compensate, but it will also obscure some detail. Further, a coating sometimes has a tendency to crack or peel.

The standard techniques fill the need for circumstances such as in research where long observation times and extremely high magnifications may be required, where the sample must be preserved for later study, and where the nature of the sample provides little, if any, alternative. However, due to the creation of artifacts and alteration of original surface features, the resultant pictures must often be interpreted; but obviously it is not always possible to be sure of precisely what changes have occurred.

My studies involve naturally existing surfaces, primarily of the "living state" of matter. It has been my intention to eliminate any interpretive uncertainty by observing and recording natural surfaces in their original condition, having a minimum of interaction and interference with the specimen. The resulting pictures show the original, virgin surfaces of the subjects; without freezing, drying, fixing, staining, or conductive coatings.

The absence of a coating frequently provides additional visual and scientific data as to the variance of elemental or chemical composition of the sample's surface. This, due to the fact that different elements and compounds have differing secondary electron emissions as opposed to the relatively uniform emission of a gold layer.

Now, although observing uncoated specimens is not entirely unique, my photographic studies are done almost exclusively with living subjects, thus, I do not have the ease and convenience of conventional observation methods. There exists a special set of problems which require special solutions.

Basically, my method requires that I apply the electron beam with great gentleness to the surface of my subjects, utilizing their natural conductivity somewhat by using a lower than normal" energy electron beam and by adjusting the instrumentation for greatest sensitivity. Actually, each specimen may require a different setting of the controls, so it is a matter of rapport as well as technique. The preservation of natural surfaces requires it's own special techniques of handling, mounting, vacuum conditioning, and fast work.

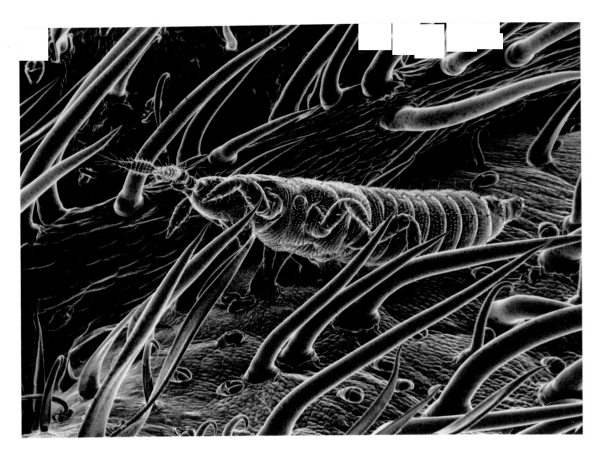

A thrip on a bean leaf, alive and uncoated (my technique).

A thrip on a bean leaf (standard technique). Same sample and area as in first photo—after drying and gold coating *Note:* It is amazing that the thrip remained on the leaf after the drying and gold coating processes.

Technical Details

**About 1 × Optical photograph
(for comparison)**

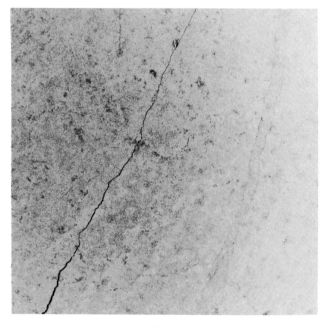

20 × "Crack in an eggshell" (gold coated)

The scanning microscopes which I use have an ultimate resolution of 100Å or better and a scan density of at least 2,000 lines per frame. The instruments are equipped with "dynamic focusing" which allows the focus plane to be tilted in order to follow the sample's surface; combined with the inherent great depth-of-focus of the S.E.M., this serves to create pictures of startling clarity.

My electron beam voltages vary from 800V to greater than 30,000V, depending upon the particular instrument used and the nature of the material being investigated. Most of the work contained in this book was done with a 5,000V beam.

While higher voltages can theoretically provide better ultimate resolution—because the electrons are of smaller wavelength, have less aberration, and can thus be focused to a finer point—it is not quite that simple in practice because the higher energy also substantially increases the penetration of the beam into the sample. Hence, when we use a high voltage, we are imaging features from further below the surface than with a low voltage beam. Now, depending on the sample material, this can actually lead to a loss of detail. The less dense the material, the more it will be penetrated. Biological materials are subject to greater penetration than materials such as metals, because of their lower density; they are also more subject to damage from burning due to the higher beam energy dissipated into and onto their surface.

For a surface feature to be imaged clearly, the volume of penetration and the size of the focused spot must be appreciably smaller than that feature, otherwise it will merely be "averaged" with adjacent features and those from below. (Refer to the closeup in the "detector and chamber" illustration.)

In actual practice, one must find the optimum overlap area for several simultaneous variables such as those mentioned for beam voltage. Other considerations are: Resolution versus depth-of-focus, for the selection of final aperture size and "working distance" to the subject; resolution versus signal-to-noise ratio, for condenser lens current—condenser current and beam voltage are factors which determine the final spot size of

70× "Crack in an eggshell" (gold coated)

252× "Crack in an eggshell" (gold coated)

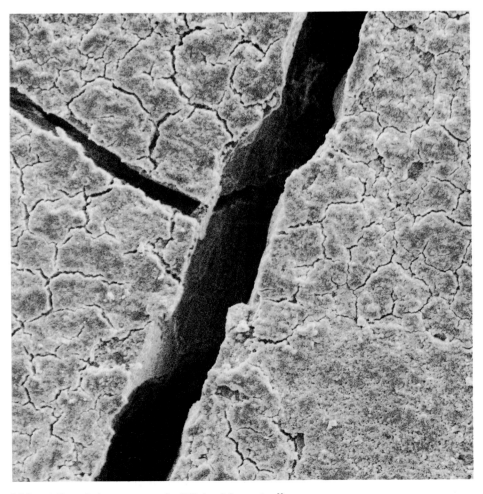

932× "Crack in an eggshell" (gold coated)

11

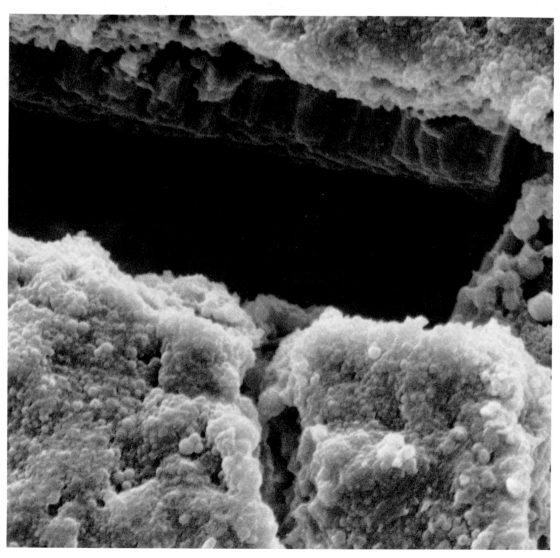

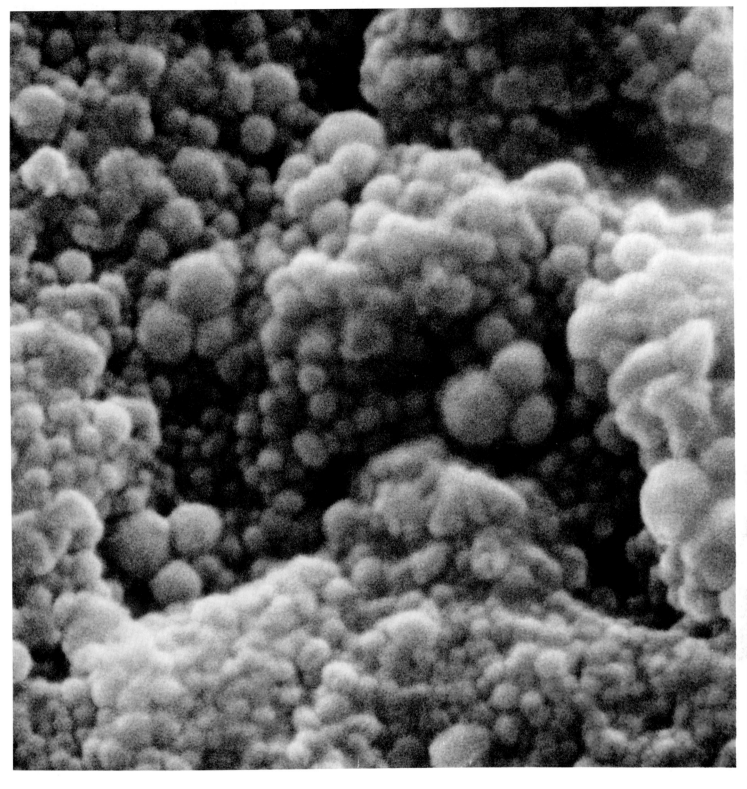

55,272× "Crack in an eggshell" (gold coated)

the focused beam. These factors must be experimentally determined with regard to the nature of the material being observed and the purpose of the investigation. A multi-level interrelationship also exists between scan line density (optical resolution), scan time, signal-to-noise ratio, electron beam current (or "beam brightness"), and the type and design of electron gun used.

I always try to keep the electronic resolution a trifle better than is actually needed to reveal the smallest detail that the optics can "see". This provides for full utilization of the maximum "dynamic range" of the instrument.

Aesthetic considerations were the final determining factor for the selection of focus depth and focus point, as well as for the "lighting" and contrast effects in the photographs.

PHOTOGRAPHIC TECHNIQUE

The established good practices of copy photography, film processing, and printing are applied.

Film formats from 35mm to 4 x 5 are used, but with the smaller film sizes, obtaining optimum resolution and sharpness becomes more critical. Also, since large murals are often produced of my work, I tend to prefer the 4 x 5 size sheet film.

In reproducing an illuminated screen, the lens is of primary importance. A copy lens with low flare, flat field, and high resolution is desirable, especially if great enlargements are to be made. In this application we are copying or time-exposing while an illuminated dot scans across a cathode ray tube. In essence, the picture is being written upon the film emulsion. It is surprising, but many an expensive scanning electron microscope is equipped with an inexpensive oscilliscope copying lens—generally adequate, but hardly superior. I use a custom lens.

A highly critical focus is essential, and this begins, optically, with the cathode ray tube (C.R.T. or T.V. screen) which is being photographed. It is a fine-grain, short persistence phosphor on a four inch diameter, flat faced screen capable of resolving 2,500 lines. Since the C.R.T. beam writes in an arc and the screen is flat, the focus point must be varied electronically in order to maintain a sharp image—"dynamic focusing" again. Consequently, the instruments usually have two C.R.T. focus controls: a center focus and an edge focus. These are "tuned" using a dot format displayed on the screen with the aid of a low power optical microscope; the controls are usually adjusted alternately several times for best results. Next, a critical focus is obtained on the film plane. This is always checked by a close examination of the negatives.

The lens f-stop is selected with respect to: lens design, which generally places its acceptable resolution limits between f4 and f11; the film speed (ASA rating); and the saturation characteristics of the C.R.T. phosphor. I begin by first selecting a film with the desired grain structure, resolution, latitude, and contrast characertistics as used with a particular developer. I try to stay on the upper part of the lens' acceptable f-range —like f8 or so—in order to maintain good focus on the edges. Lastly, the brightness of the recording C.R.T. must not be excessive or the picture's tonal range will be limited due to saturation of the phosphor; also, internal reflections from the C.R.T. face, the lens, and other inner camera structures will probably cause an unacceptably high film fog level. Obviously, one must find the optimum balance of these factors for the particular S.E.M. being used.

Kodak's Plus-X (4147) is probably my most used film, with Ektapan (4162), Commercial blue-sensitive film (4127), Panatomic-X for smaller formats, and Polaroid's PN55 and 665 positive-negative films also being used often. Polaroid's type 52 is also useful in circumstances where no negative is needed. The Polaroid products, of course, have an obvious advantage in their time-saving ability— especially in experiment and test procedures where the result of one test determines the conditions for the next. The (4127) blue-sensitive film has an advantage in ease of handling, as one may load and process under a safelight; however, for normal work, its latitude is not as great as with the other films mentioned, and the choice of developers is limited. In certain applications though, the higher contrast of 4127, as used with the appropriate developer, makes it an ideal choice.

The developers which I use for the films mentioned are: Kodak's D-76, Microdol-X, DK-50, and HC-110—for normal work; and D-19 or D-11 for high contrast work.

For printmaking, I have used a Beseler 45MCRX condenser enlarger equipped with a 135mm, f5.6 Schneider Componon lense.

A JOURNEY INTO MICROSPACE

For me, the experience begins out in the fields walking beneath the open sky where I come across likely subjects for my studies.

My finds are transported to the laboratory with great care. They are protected from dust, dirt, and damage, as I require virgin surfaces, free from injury of any kind. A sample is carefully selected from the ones gathered. I look at it closely with the naked eye or a magnifying glass, whereupon I can immediately view that first order or magnitude; "smaller than we usually bother to notice." The sample is placed in the vacuum chamber of the S.E.M. on a stage whose mount allows it to be manipulated with five separate motions: x axis, y axis, z axis, tilt, and rotation. The chamber door is sealed and the pumpdown begins, reducing the chamber pressure to about a millionth of an atmosphere in a few minutes. The lab lights are turned off and I take my place at the controls. Power to the electron gun and sensors is turned on; preliminary checks and adjustments are made; and I set the controls for the first order of magnification: 5X to 100X.

After some tuning, the picture appears, and at the lowest power, I can see more clearly, but differently, what I saw with my naked eye. In that first view after the object is in focus, at only 10X or 20X, Nature and the way the electron beam illuminates the object have already captured my full attention and started a flow of almost hallucinatory images. A few degrees rotation of the zoom control and I am already beyond the realm of our normal senses. It is easy to dwell on this level, but I go on, further in. The next switch takes me to a magnification range of 100X to more than 1,000X. Now, I find myself looking at the details of my first finds on the subject's surface. I see small hairs upon larger hairs and many things that have no easy names. Many of these images are formed in the eyes and in the mind by the way that things seem to magically line up. Of course, the more "real" objects we find have their own kind of beauty. Besides the beautiful images they evoke, they take the mind to other speculation—a specific kind of wonder—knowing that they are, indeed, *real*.

By now I am doing some real traveling—through jungles, over hills, into valleys. . . . Intuitively, my fingers drive the controls; the mechanical movement with one hand and all the rest of the controls with the other—zoom, focus, attitude, brightness, contrast, etc.—about a hundred in all. I glide over the "landscape", getting closer and closer but never touching. Climbing the tongue of a fly to glide slowly over its almost perfectly faceted eye; taking the time to go down among some of the individual facets and explore the occasional hairs in-between; and even closer, to study the grooves and texture of the hairs themselves. Or, examining the surface of a Yellow Tarweed flower, I have come across strange little inhabitants with antennae that looked as though they had been turned on a tablemaker's lathe—thrips, who, wandering about the petal scattering pollen, were frozen in their tracks when the air was removed from the chamber. In another trip, I gazed in awe at the topography of a female marijuana flower, ripe with resin nodules (the sacs which contain the pure essence of hashish) looking like tiny "beings", each one having a different personality in my imagination. After spending much time at this level, I decide to try the next: 1,000X to 10,000X. Here, the controls become very critical and require a great deal more sensitivity to operate. Also, this level, while rich with many features, is not the cornucopia that I found on the first two levels. I can see the wrinkles and texture of a cell wall among a sea of thousands of other cells, even the bumps on the little hairs that reside upon larger hairs. But not much is new, I'm just a bit closer to what I saw before.

The higher I go in magnification, the more barren the surface seems to get. The resolution limit of the S.E.M. is becoming evident. Finally, as I go to the last order of magnification—10,000X to 100,000X—everything is so sensitive that it is difficult to even touch the mechanical controls without losing my place; I move around by controls which are purely electronic. Here, I must know precisely what I'm looking for, because obtaining a clear image requires that much time and great care be taken to tune the

S.E.M. for each of several variable operations. But even with pinpoint tuning, the image at 20,000X and higher is comparatively fuzzy and lacking the richness of detail to be found at the lower magnifications—a combination of the instrument's limitations and the fact that nature just didn't seem to put quite as much of what I am seeking at this niche in the continuum of physical space—at least not for this method of seeking.

Of course, I have been stopping at several places along the way to take photographs; it takes about seventy seconds for each frame. The decision of when to stop and record can be a difficult one: I think that I should look everything over before taking any pictures, but I may not find what I initially saw when I return, or it may have deteriorated due to the drying effect of the vacuum. Yet, it is always possible that the most exciting find may be just over the horizon. It is difficult too, to keep from being totally absorbed by the wonderful sensation of movement in the exploration itself. However, the process turns out to be a most natural one, like floating down a river with many courses: I have only to decide which one to take.

The next step of the journey takes me to the darkroom, where the momentum of excitement continues. The laboratory shooting only provides a teaser of what the final image will be, because the S.E.M.'s viewing screen is capable of far less resolution than what may be expected in a final print. When the negatives are developed, the mystery still prevails, as they provide only a reversed image that is difficult to interpret. But now, I can at least get a clearer idea of what are the best possibilities for printing. Using an optical microscope, I examine the negatives to judge the limits of their sharpness. Many times, this is a startling visual experience in itself, for when the picture is sharp and the density of information high, the images begin to emerge before my eyes once more, but in even greater relief. Next, I make contact prints for a look at the positive picture. I can now see things that were not apparent even in the negative. And finally, I know which ones will make it to the final stage.

My instincts as a photographer take over completely as I continue. The contact print has supplied just enough information to tell me what the final print should be like. I make tests and adjustments for cropping, emphasis, contrast, and brilliance; press the enlarger switch; expose; and then immerse the exposed paper in the developer bath. Even then, I can't keep my hands off the image, using their warmth to assist the development of particularly tricky areas of the print.

When the darkroom lights are turned on, I see the final visual record of my journey. Inconceivable in the field, not imaginable in the laboratory in this way, here is an object, independent, to be viewed and shared with others. Yet, this by no means challenges or diminishes, for me, the importance of the events that have led up to it. It does serve, in my mind, as a bond and lasting memory of the whole experience. Although I do have the motion and changing views now, I do not have this artifact, which may be examined at great length, for the richness of it's detail, for the unique chiaroscuro and imagery of the electron beam process, and lastly, for the knowledge provided about the world's smaller inhabitants and structures.

INSECT LIFE

You are about to view some of the earth's most successful inhabitants—the insects. Having been here for over 400 million years, they account for over three-fourths of the world's animal species; over 750,000 species are known and described but there remain many more to be classified, perhaps as many as 5 million. The present population of insects in the world is estimated at somewhere around a million trillion (10^{18}); and insects, not only vastly outnumber human beings, they *outweigh* us by about 10 to 1!

From at least one point of view, some of the relationship between man and insect is manifested as destructive, as with food infestation, crop destruction, stinging and biting, and spreading disease—not to mention spreading insecticide. Some of the interaction is mutually beneficial: It's wonderful to have honey to eat and for our fruit crops to be pollenated. The honey bee, of course, is responsible for both.

Spiders have an interesting place in the scene. By catching stragglers and the unwary they check some of the growth of the insect population, and thus contribute to the preservation of balance. Mites are very minute—about the size of a speck of dust—and difficult to study. Even so, some 25,000 species have already been identified; and this is estimated to be only about 10% of the total. Some species of mites are destructive—ruining crops and being parasites—while others are helpful—preying on more harmful mites and insects. Spiders and mites are arachnids.

Although the specimens of insects photographed in this section were gathered from the Southern California area, these little animals can be found in most backyards. They are common insects and are generally present throughout much of the world.

The following photographs are of some of the more interesting specimens that I have investigated in the last few years; they are in no way intended to be a formal study of insect life or structure.

Usual S.E.M. preparation techniques kill the specimens before their examination. In this case, great care was exercised to keep subjects alive and well for the shooting. Consequently, some of the specimens were returned to the garden alive. The most extraordinary example was the jumping spider who survived over 30 minutes of high vacuum and high voltage.

A question often asked is how one gets an insect to hold still while scanning him for over a minute to obtain a picture. Well, the truth is that they do not all hold still; although the high vacuum tends to immobilize them, many a good photograph has been ruined because of an unpredicted movement. This effect is multiplied at very high magnification where the mere heartbeat of a small animal can cause enough vibration of a limb to make photographing impractical if not impossible.

400 x

A Fly's Eye
Lesser Housefly (*Fannia Cannicularis*—Family Muscidae)
A compound eye. Each facet is part of an individual eye (ommatidium) that has its own lens and retina. The fly's brain integrates the many images of the individual eyes into one, "understandable" image.

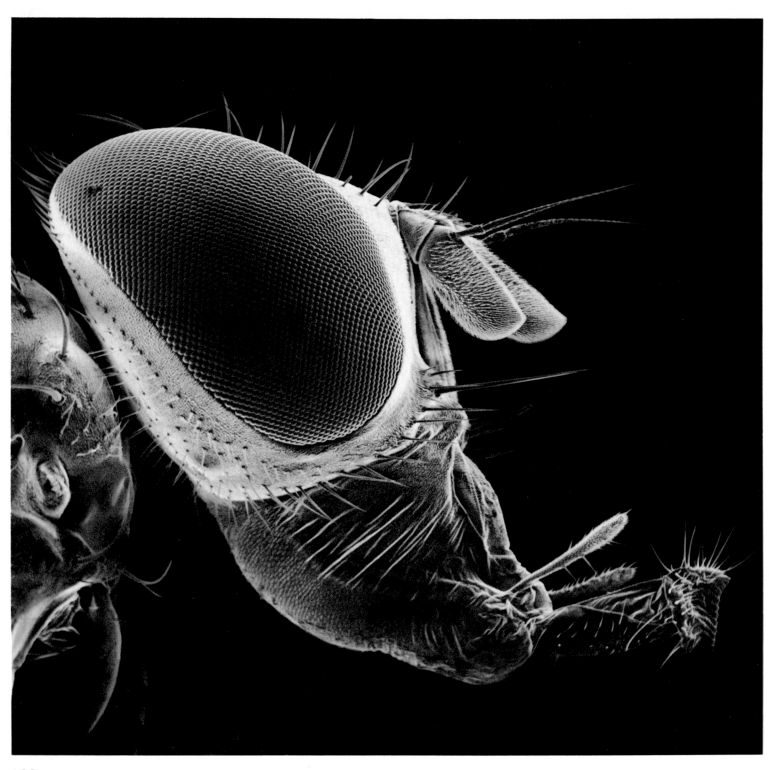

100 x

Housefly Portrait
Lesser Housefly (*Fannia Cannicularis*—Family Muscidae)
A very common fly. The extended proboscis is hinged and is normally stored "up" when not in use.

Fly's Proboscis
Blowfly, Shiny Green Garbage Fly (*Family Calliphoridae*)
Labellum underside; The pseudotrachea, or lines of holes, in the sponging disc
channel to the center hollow tube.

300 x

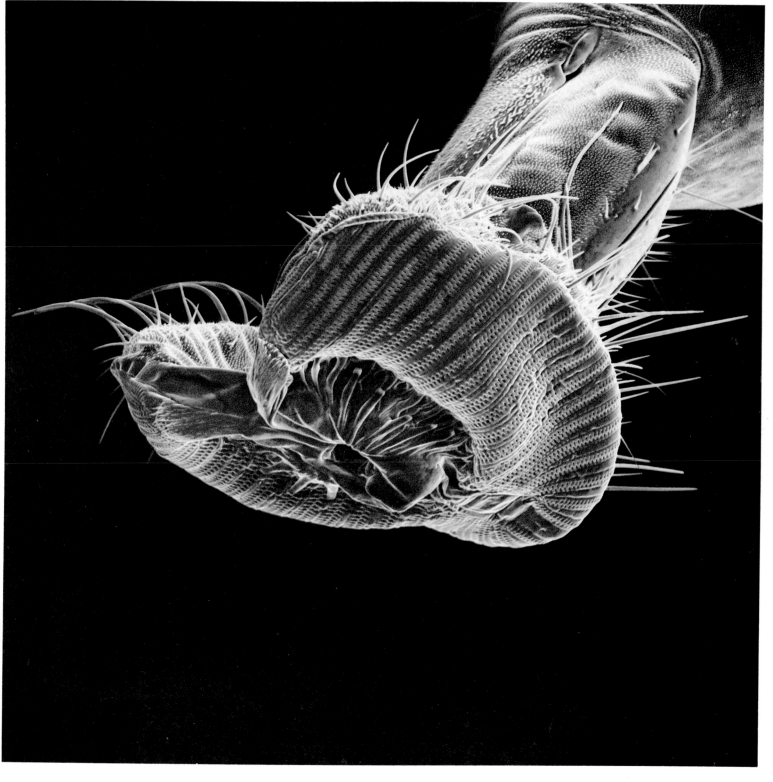

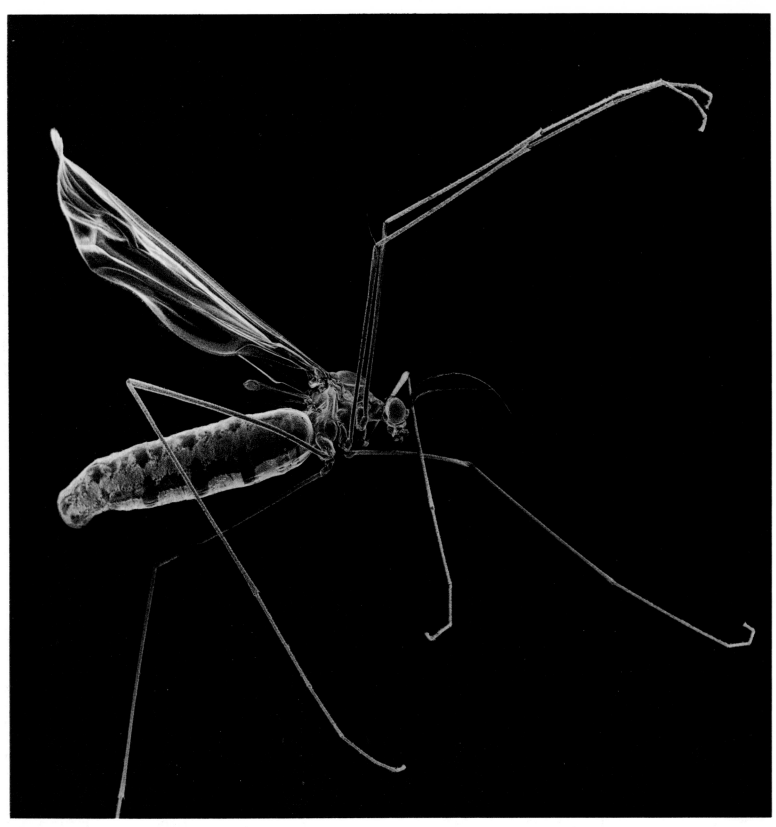

27 x

Crane Fly
(Family Trichoceridae) (side view)
This particular type of Crane Fly is relatively small. It resembles a mosquito and is about the same size.

"Mister Woolly"
Whitefly (Family Aleyrodidae) (front view)
A minute white fly found on plants; he is covered with waxy flakes (1 mm size)

725 x

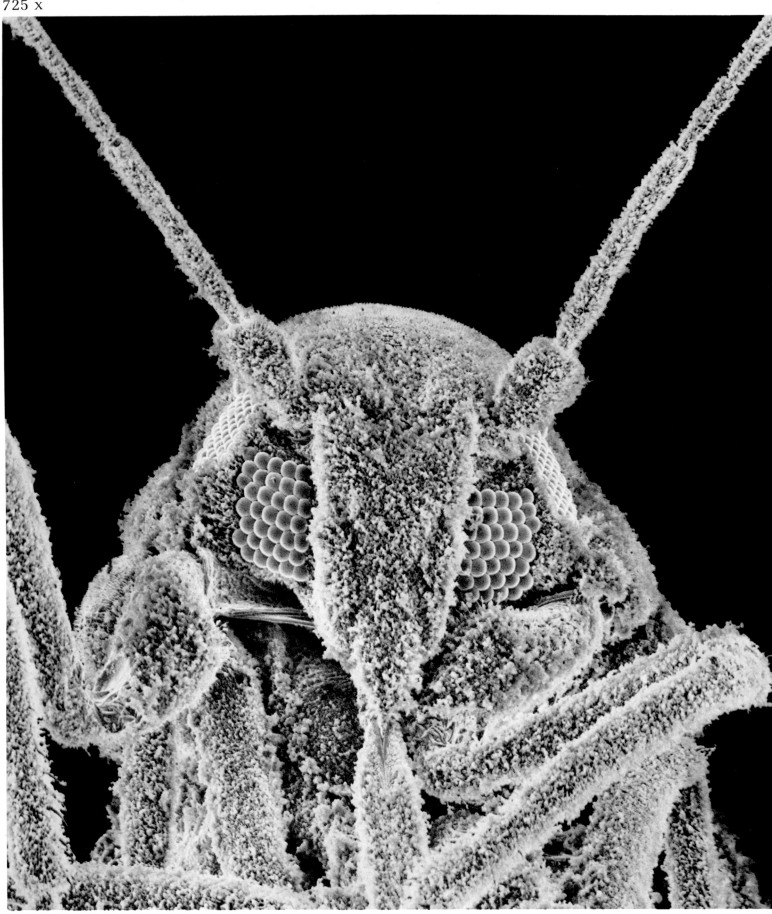

44 x

"Portrait in Space"
A Diamondback Moth (Family Plutellidae)
This is a small moth, about ¼″ long which sometimes infests food, grains and
fabrics. (Object in background is the specimen mounting pedestal.)

50 x

Jumping Spider
(Order Araneida) (*Phidippus*—Family Salticidae)
It is said that these spiders have about the best vision among invertebrates. Note eight eyes.

Spider's Foot
Tarsal claws of Jumping Spider (*Phidippus*)

165 x

A Treehopper Nymph
(Family Membracidae)
This insect damages plants by sucking the juices.

48 x

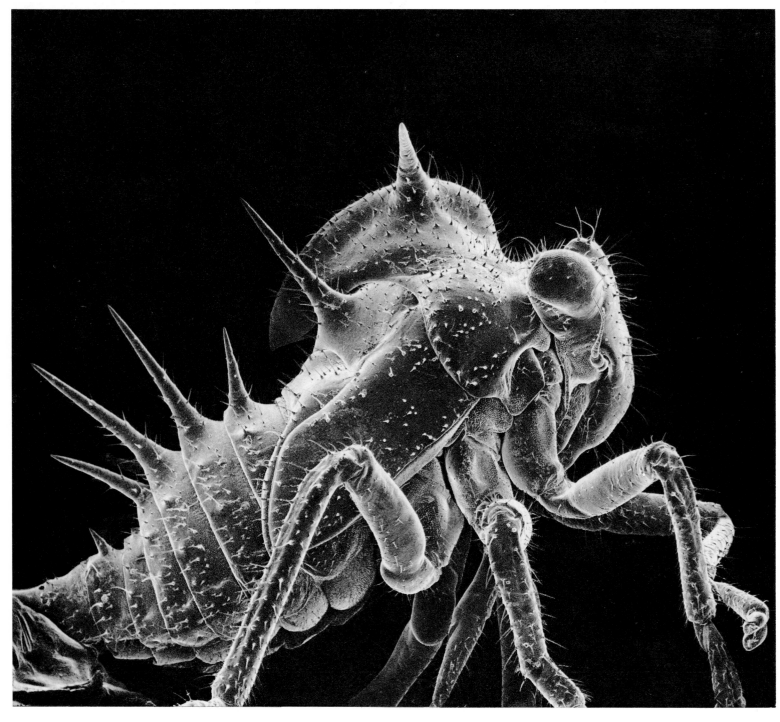

27 x

Portrait of a Vespid Wasp
Yellow Paper Wasp (*Polistes fuscatus aurifer*) (Family Vespidae—Subfamily
Polistinae)
This is a large, very common wasp, abundant in summertime.

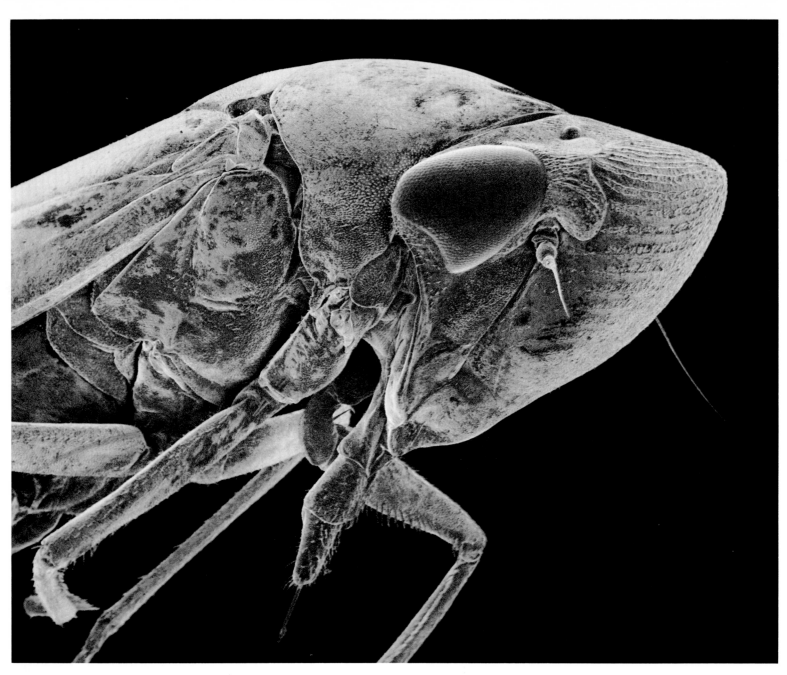

63 x

"Blue Sharpshooter"
Leafhopper (*Hardnia Circellata*—Family Cicadellidae)
Another insect that damages plants by sucking the juices.

"Mates"
Mediterranean Fruit Flys (*Ceratitis Capitata*—Family Tephritidae) Male-left
Female-right
(Front view) Notice the "flags" on top of the male's head. This insect is of great
agricultural importance—doing enormous damage to fruit crops in Hawaii,
California, and many other places. They lay their eggs in fruit which the larvae
eventually destroy by eating away.

52 x

52 x

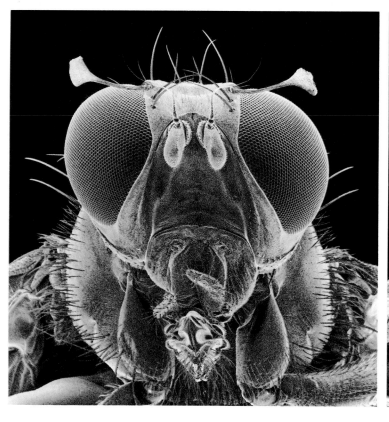
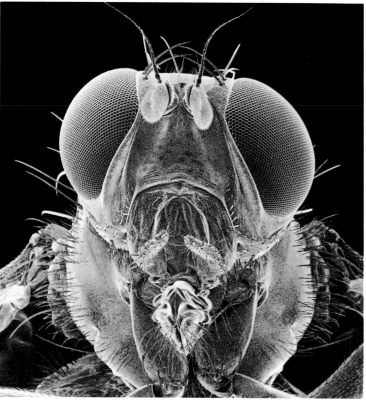

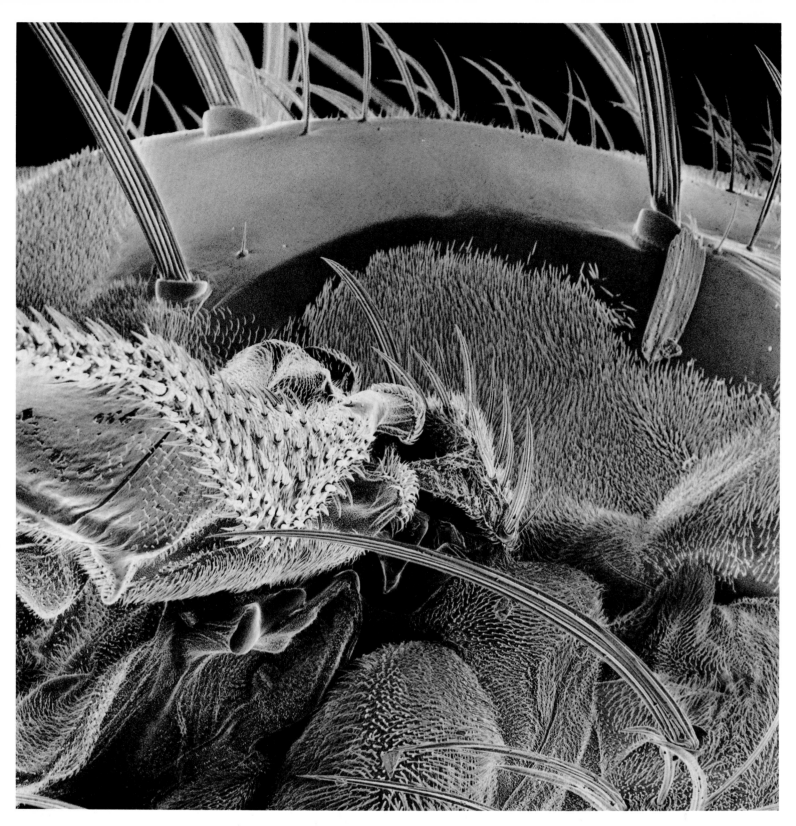

485 x

Landscape on a Fruitfly
Mediterranean Fruitfly
Structures near the wing.

"Northwest Slope"
(Family Tettigoniidae)
Surface structure of the back of a Katydid nymph.

405 x

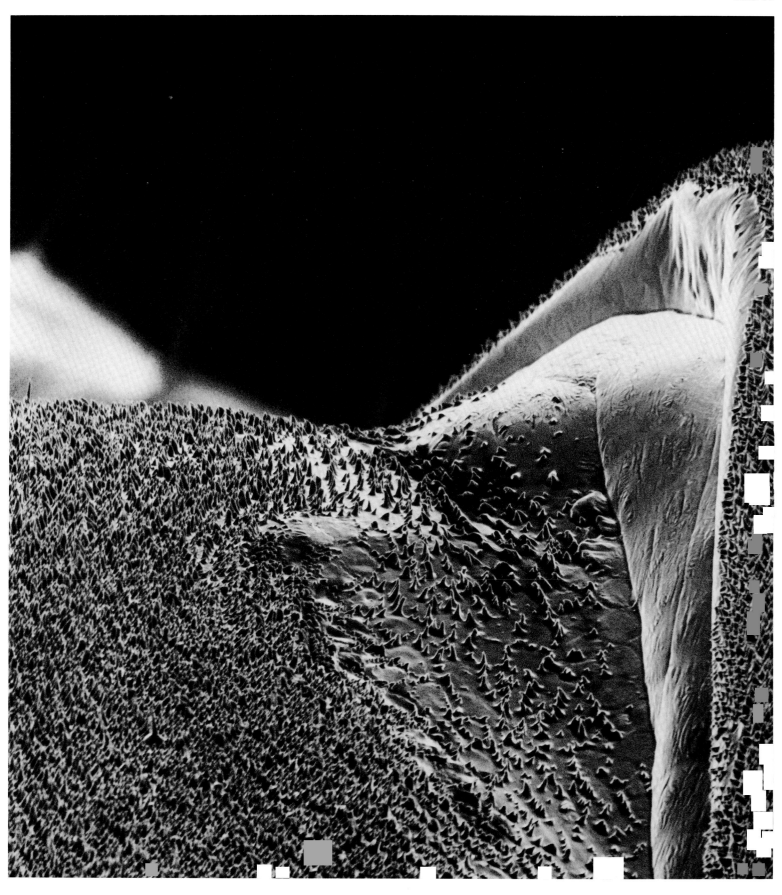

Termite

(*Reticulitermes Hesperus*—Family Rhinotermitidae) Subterranean termite
Underside view of a worker termite.

108 x

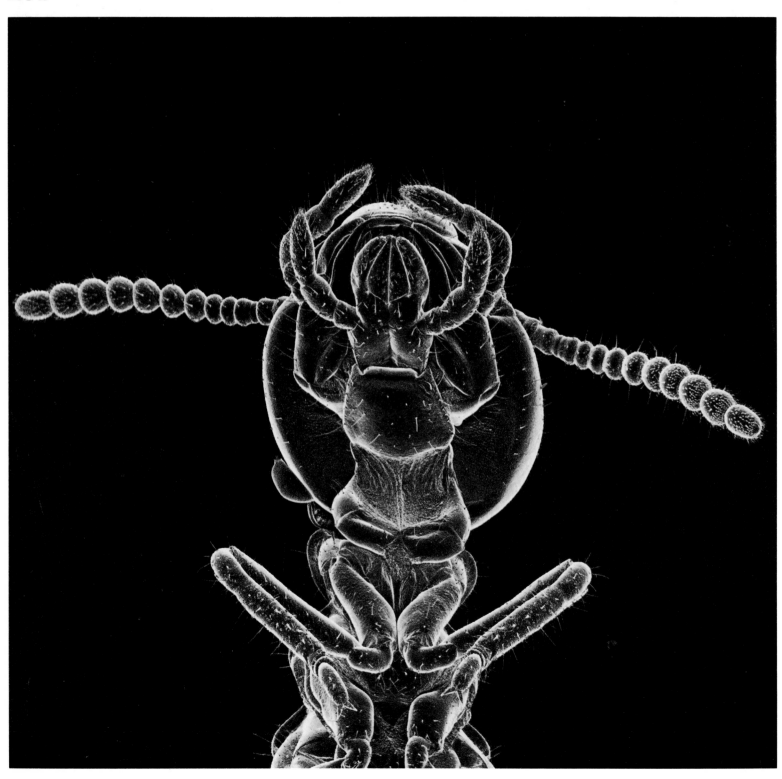

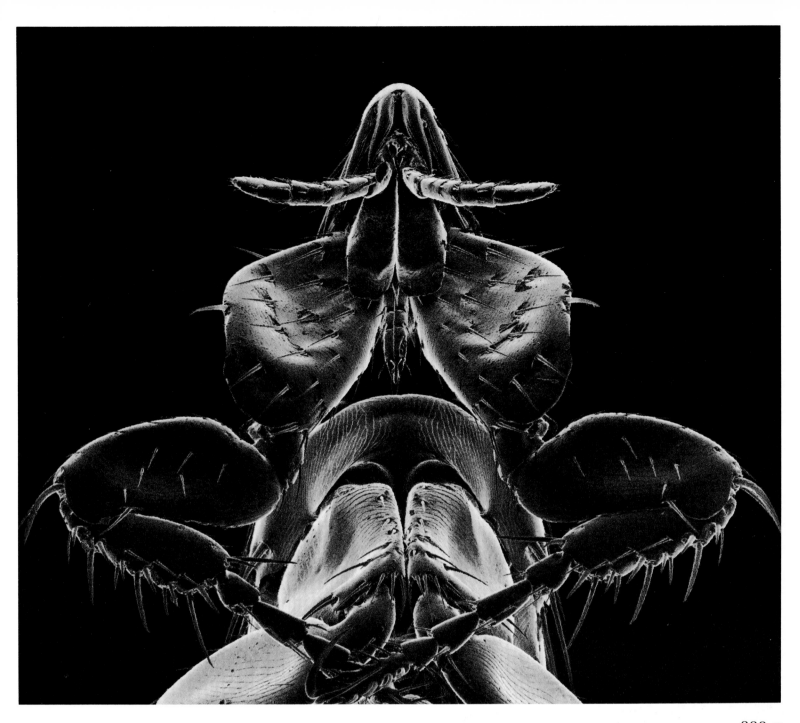

222 x

Flea
(Family Pulicidae)
Underside view of a common dog flea.

111 x

Mosquito
(Family Culcidae)
A newly emerged male mosquito, showing proboscis, maxillary palp, and antennae.

Braconid Wasp Portrait

(Family Braconidae)

A minute wasp. The actual size of this specimen was about 2½mm. This wasp is an important insect because, as a parasite of other insects, it helps to control some noxious pests.

190 x

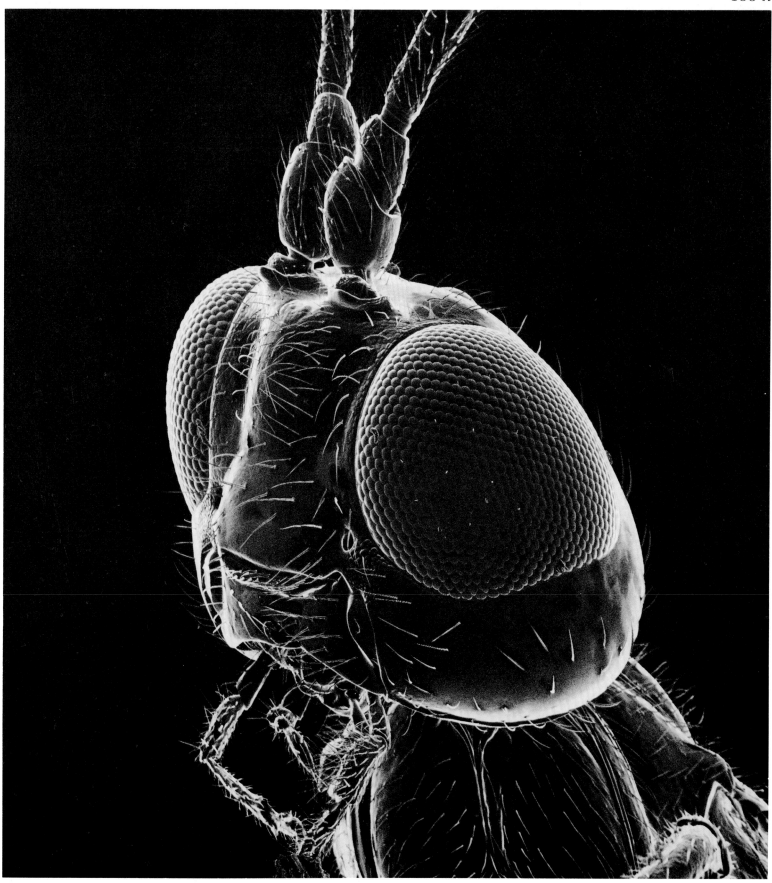

Gnat

A Gall Gnat (Family Cecidoymiidae)
A minute fly. This specimen was about 1½mm. (Front view)

401 x

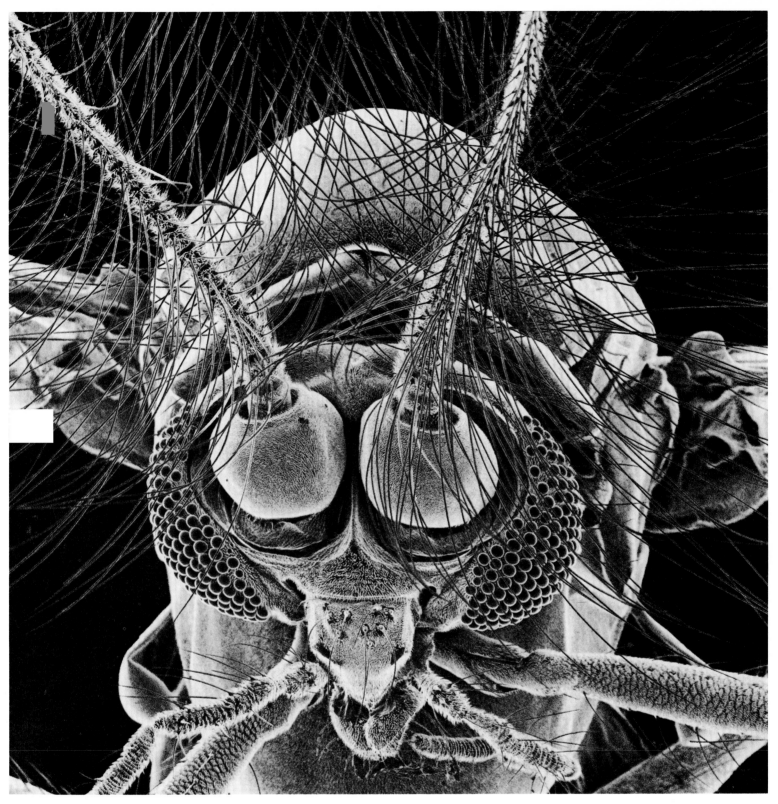

476 x

Feathery Midge
A common midge. Male. (Family Chironomidae)
The size of this insect is about 2mm. These midges are aquatic breeders and do not bite. They can be seen in large swarms in the summer.

A Mite on the Neck of a Termite
Mite (order Acarina—probably an *Orabatid* mite) Termite (Western Subterranean)
Note that there is another organism—probably a bacterium, yeast, or mold—at the upper right quadrant of the picture.

1,100 x

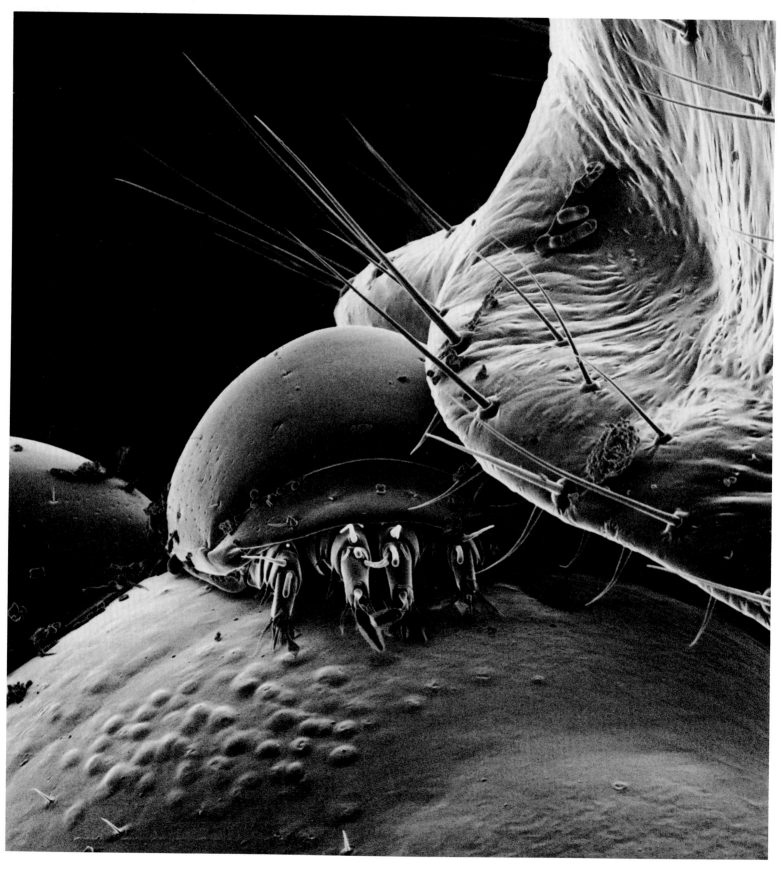

730 x

Moth Fly
(Family Psychodidae)
The size of this specimen was under 1mm., making it small enough to fly through a window screen. These cute little fellows breed in sink and bathtub drains.
(Front view)

33 x

A Honeybee Portrait
(*Apis mellifera*) (Family Apidae—Subfamily Apinae)
Worker bee. The many hairs are an adaptation to pollen gathering.

Honeybee's Eye
Compound eye. Note ocellus (simple eye) in background.

280 x

"A Beescape"
Small Carpenter Bee (Family Apidae; *Ceratina*, Subfamily Xylocopinae)
Area near wing showing part of tegula (a scalelike structure overlying the base of the wing).

615 x

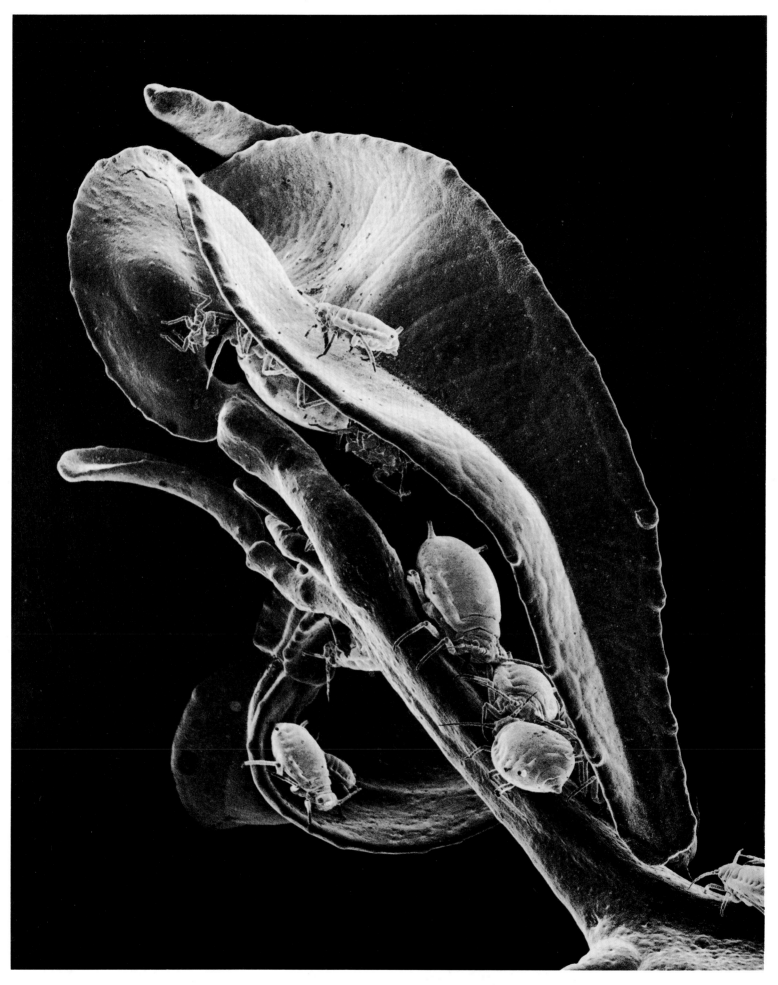

33 x

"Aphids Grazing on a Lemon Tree Leaf"
Aphids (Family Aphididae)

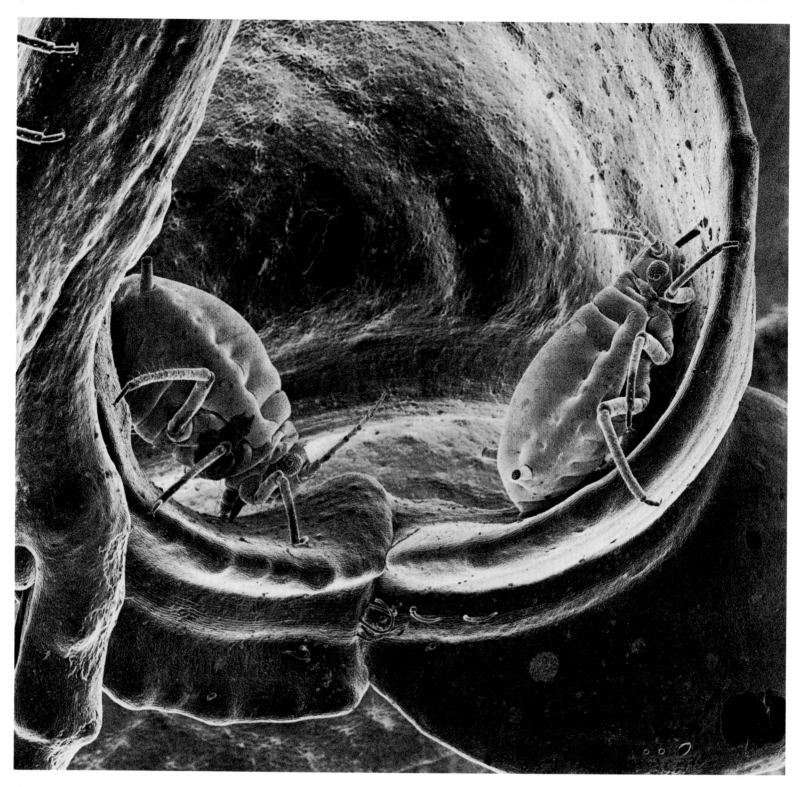

83 x

Two Aphids on a Lemon Leaf
Aphids (Family Aphididae)

Sensillae
Cluster of sensors on an aphid's antenna.

3,650 x

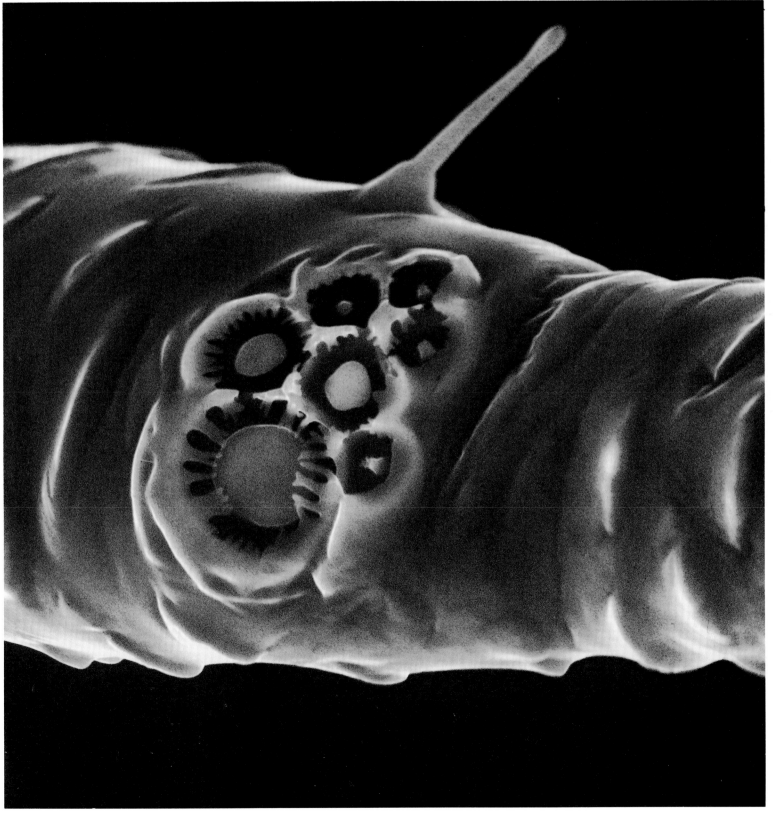

40 x

40 x

Two Harvester Ants
California Harvester Ant (*Pogonomyrmex*—Family Formicidae)
Specimen No. 1—side view; Specimen No. 2—front view.
Half inch giant, "bearded", stinging, red ant.

"Portrait of a Brown Ant"
Common Argentine ant (*Iridiomyrmex humilis*—Family Formicidae)

100 x

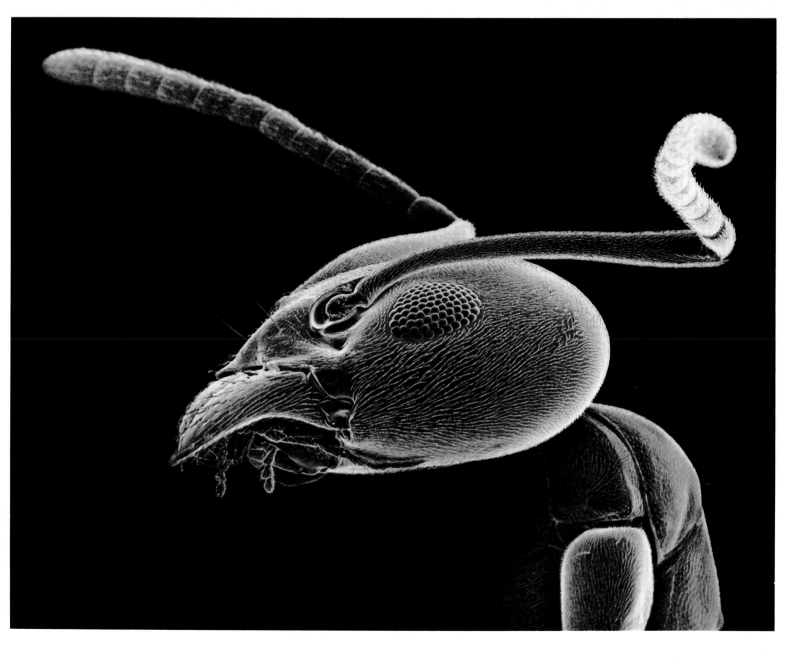

Velvety Tree Ant
(*Liometopum Occidentale*—Family Formicidae)
Furry, red and black tree ant.

188 x

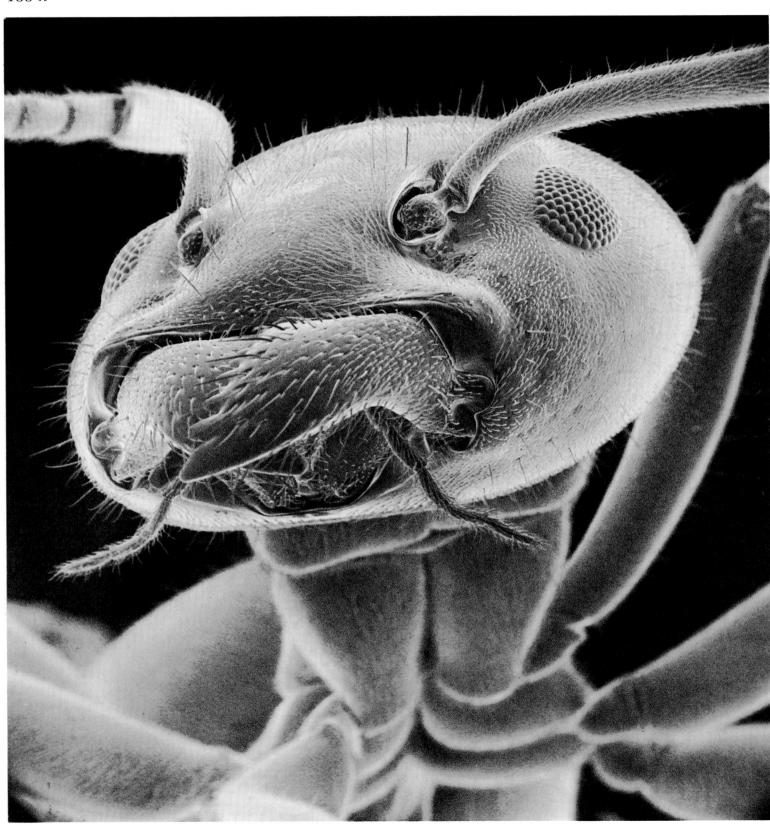

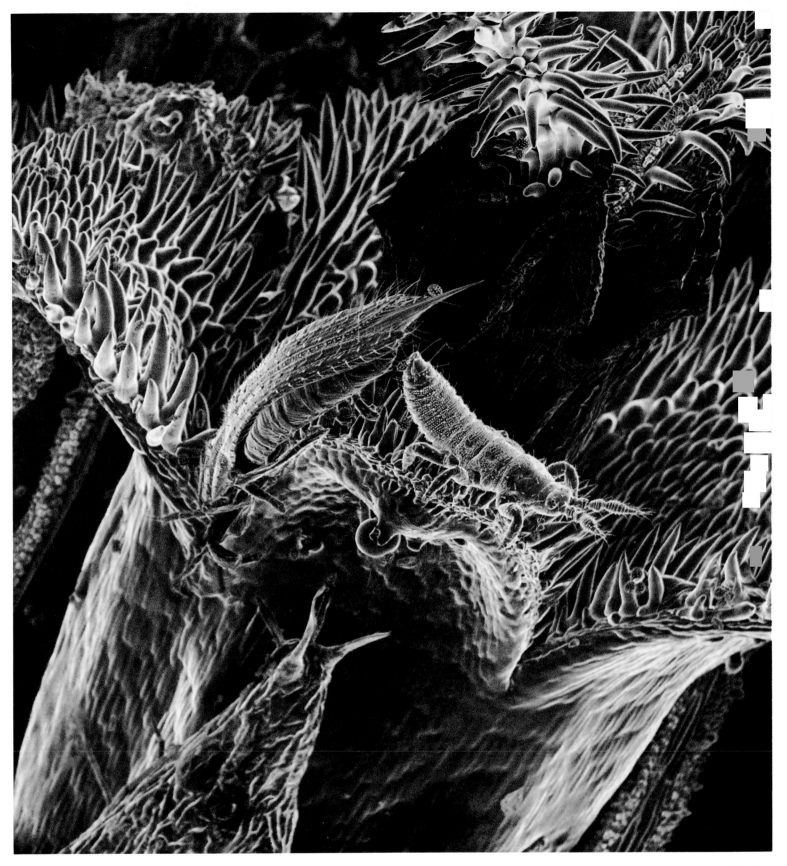

190 x

"Two Thrips (Order Thysanoptera) (Adult—left/Nymph—right) On a Yellow Tarweed Flower (Hemizonia Virgata) With Pollen Scattered About"
Thrips (Family Thripidae)
The thrips are on the petal of a disc flower; flower stigmas are upper right; note also, two kinds of pollen.

71 x

A Thrip on a Wild Flower
(Family Thripidae)
Very common small insects (1mm or so) found on plants. Sometimes plants are damaged by their feeding; some thrips are vectors for plant disease.

Thrip on a Wild Flower Closeup

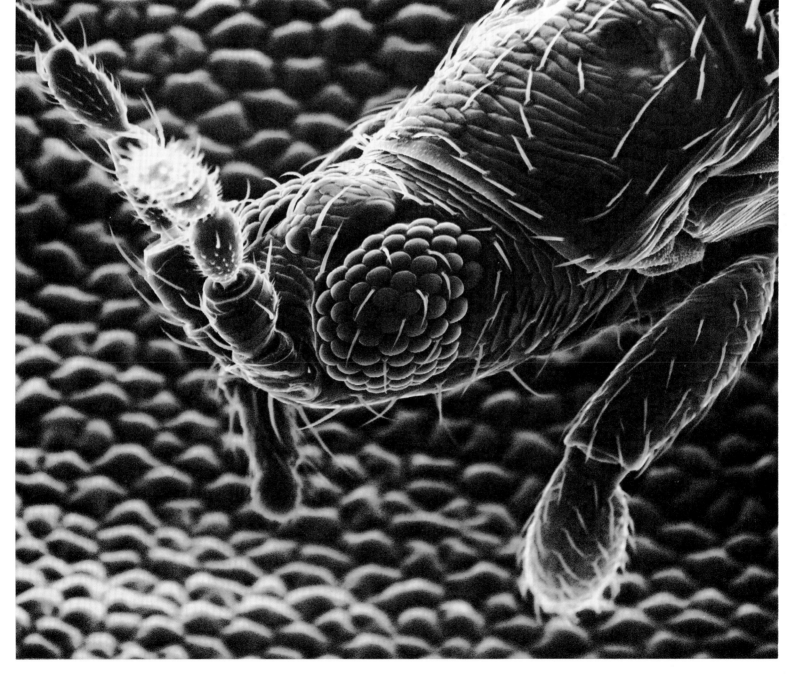

"Prowler"
(Order Acarina; Suborder Trombidiformes—Family Tetranychidae)
A Spider Mite on the underside of a Marijuana leaf. The mite has, seemingly become entrapped in the resin of a broken nodule.

730 x

Spider Mite on a Cannabis Leaf
Tetranychid mite (front view)

PLANT LIFE

The oldest plant fossil remains have been dated at around two billion years, making them a rather ancient life-form. Plants are primary receptors for solar energy, which they use to create organic matter. They are also the prime agents for the production of oxygen on earth. Both of these actions take place through photosynthesis. Animals breathe oxygen and give off carbon dioxide; plants are a mirror image in this respect because they "breathe" carbon dioxide and give off oxygen. In one way or another, animals are ultimately dependent upon plants for food.

It is estimated that there are over 350,000 kinds of plants. The photographs in this section deal with the flowering plant, or angiosperm, of which there are about 200,000 species. The origin of the flowering plant is still unsure but it is estimated to have appeared some 250 million years ago, probably in the tropics. Its evolution is closely related to that of the insect and there are as many interesting adaptations in flowers to insure pollenation by insects (of course the wind accounts for much pollenation too), as there are adaptations by insects to pollenate and feed on plants.

With the colorless but fine "light" of electron illumination, we can see plants and their flowers in a new way with great clarity and depth. Their structure and form appears as no eye could see it with ordinary light; it is further enhanced by the unique interplay of highlight and shadow. The specimens were, of course, alive and in their "natural" state—gathered from the Southern California area. Many are rather common plants found throughout the United States and in many other parts of the world.

It should be noted that while there is much useful scientific information present in these recordings, they are intended primarily as visual studies.

500 x

A Candytuft Flower Stigma
(*Iberis*—Family Cruciferae)
Stigma (female organ) impregnated with pollen.

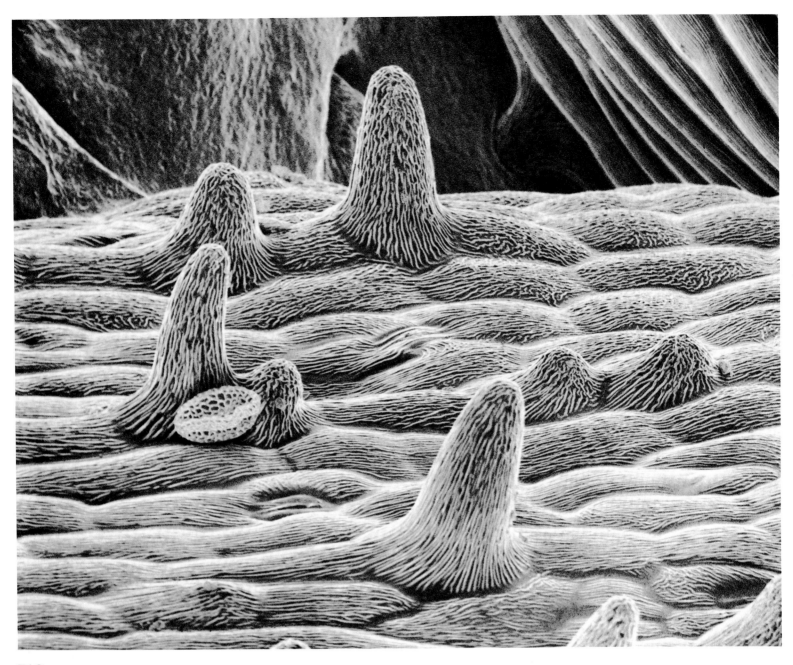

710 x

Candytuft Flower Stem-Scape
(*Iberis*—Family Cruciferae)
Note piece of pollen next to lobe on left; stomata (plant respiration pore) in front
of it and another in the center of the picture.

Candytuft Flower Petal

(Iberis—Family Cruciferae)
Cells at a crease.

1,175 x

Oleander
(*Nerium Oleander*—Family Apocynacea)
Stigmas and petal.

21 x

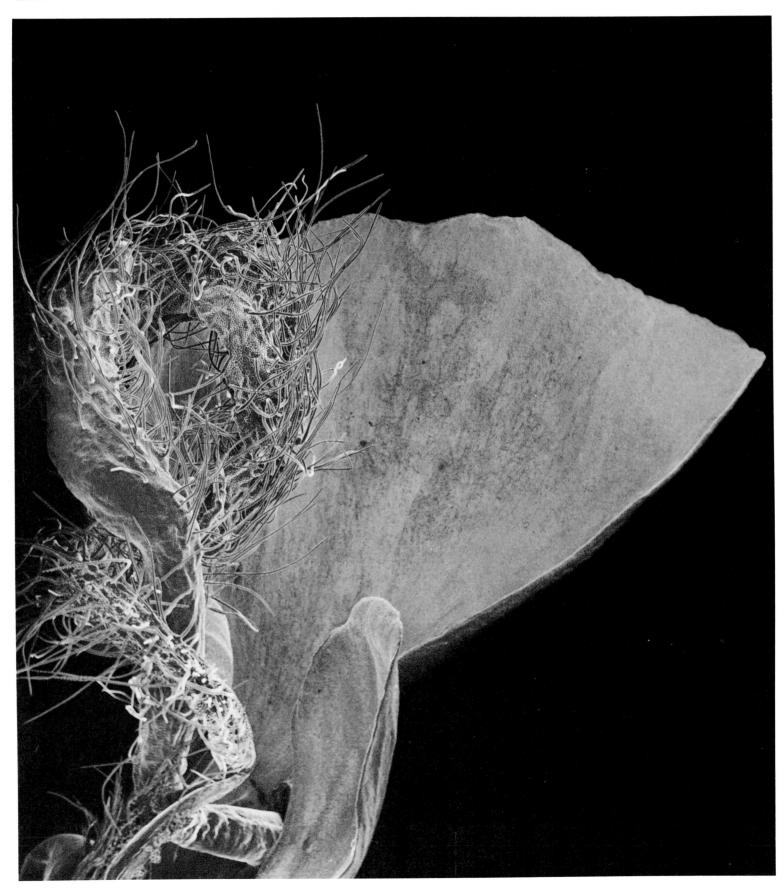

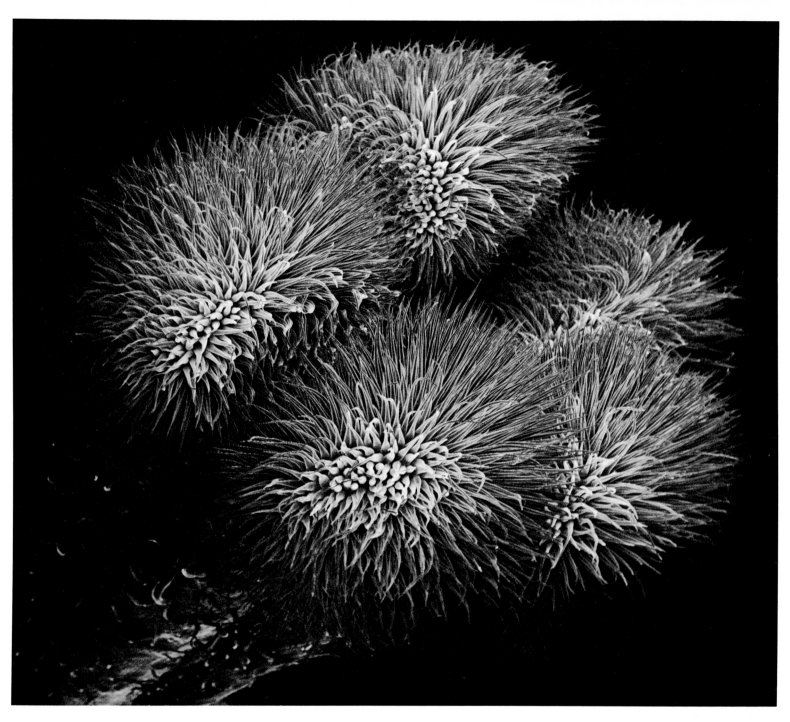

44 x

Hibiscus Pistils
(Family Malvaceae)
Feathery stigmas of the pistils. The pistil is the female organ of the plant and the stigma is the part of the pistil on which pollen is received and germinated.

Carolina Jasmine
(*Gelsemium*—Family Loganiaceae)
Style and anther—filled with pollen. The style is the stem part of the pistil and the anther is that part of the male organ which bears the pollen.

55 x

Gladiola Pistil
(*Gladiolus*—Family Iridaceae)

56 x

"Geranium Triplet"
(Family Geraniaceae)
Three views of one Geranium pistil.

"Friends"
Strawberry flower (*Fragaria*—Family Rosaceae)
Immature fruit showing achenes (the oval seed-like fruits which cover the strawberry) with stigmas and styles protruding.

80 x

Castor Bean and Flower
(*Ricinus communis*—Family Euphorbiaceae)
Female flower and spicate epidermis (spiney hairs) of the bean fruit.

68 x

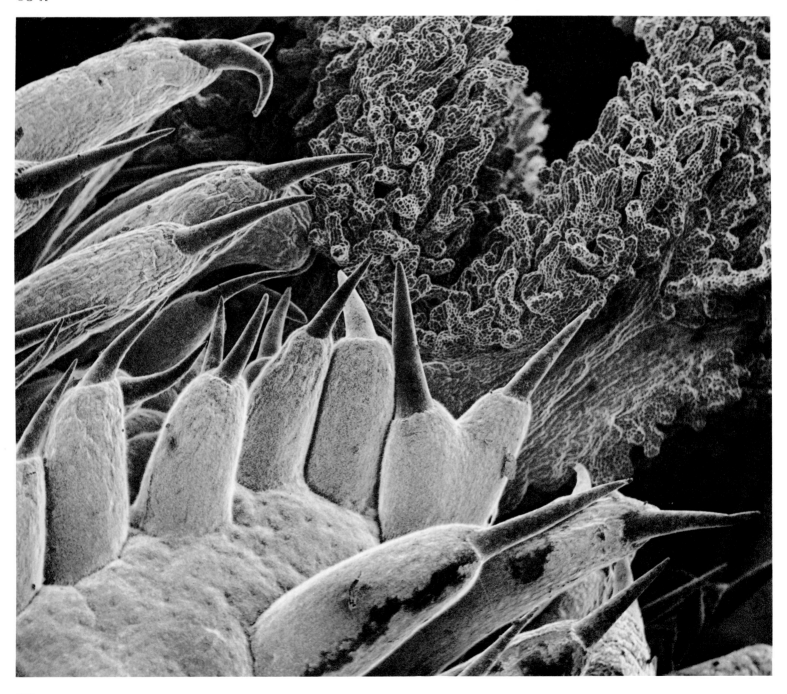

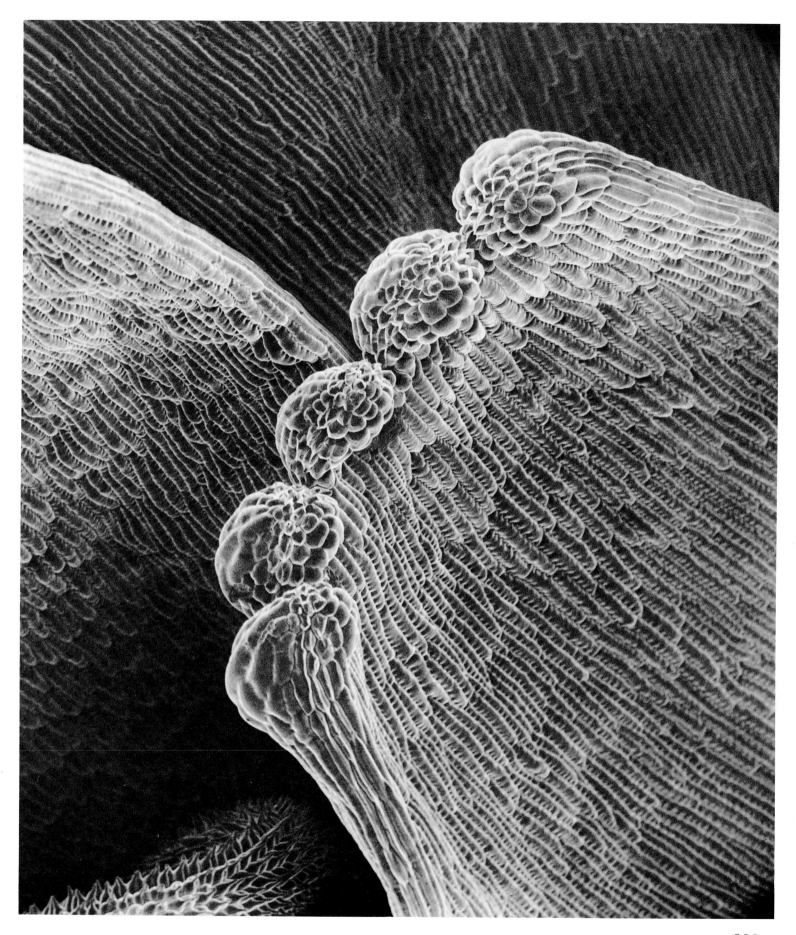

300 x

"Dandelion Composition"
(*Taraxacum vulgare*—Family Compositae)
Common Dandelion, petals and stigma.

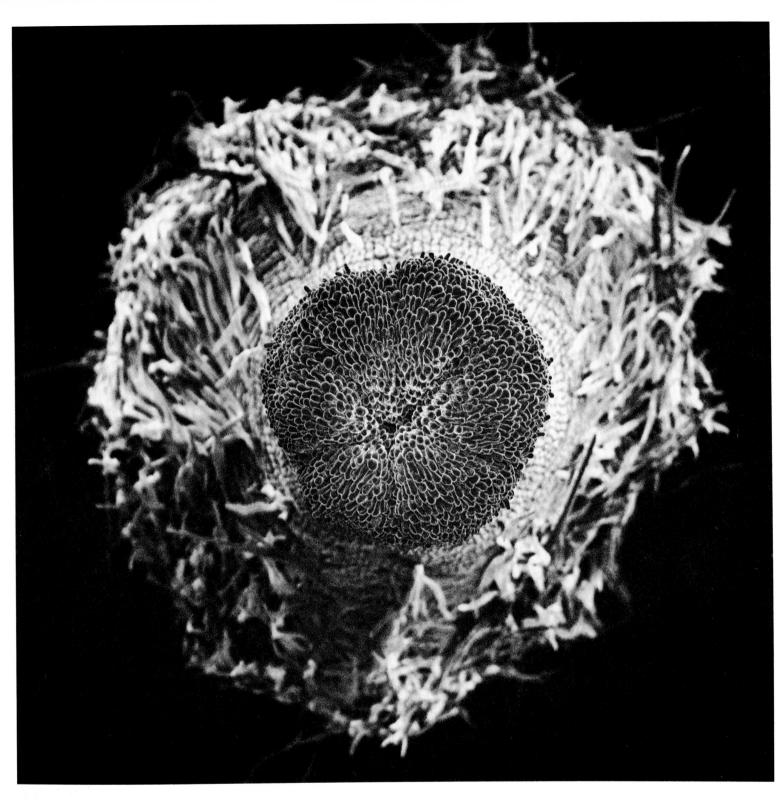

125 x

Victoria Spring Tree Flower Pistil
(*Pittosporum*—Family Pittosporaceae)

Narcissus
Daffodil (Family Amaryllidaceae)
Stigmatic surface of pistil.

106 x

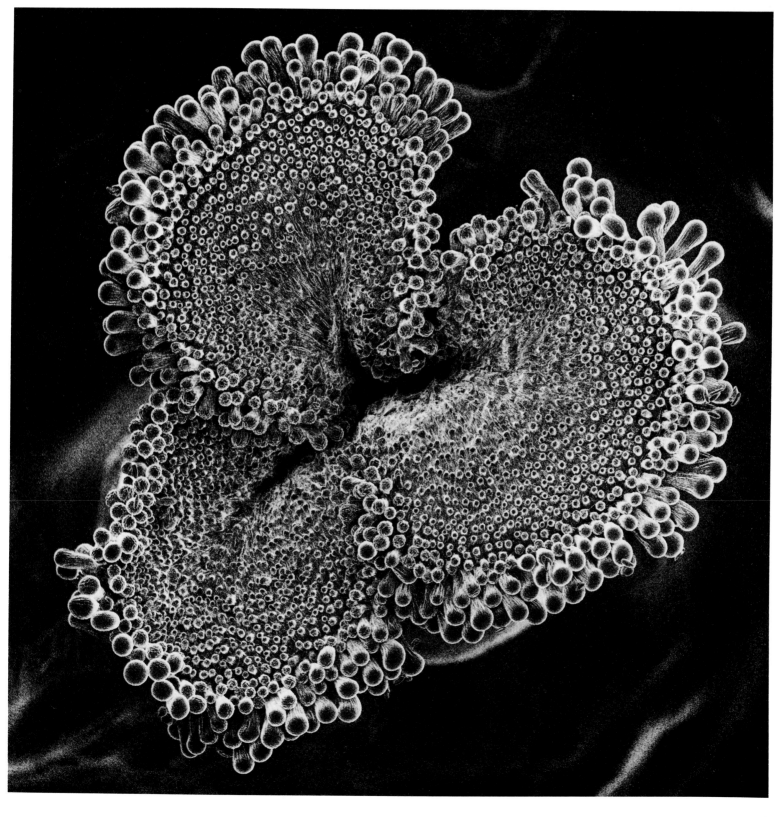

Curved Structure

Curved protuberance from the style of a Hibiscus flower, enclosing a multicellular trichome. A trichome is a type of plant hair—sometimes they are glandular and secrete plant resin.

144 x

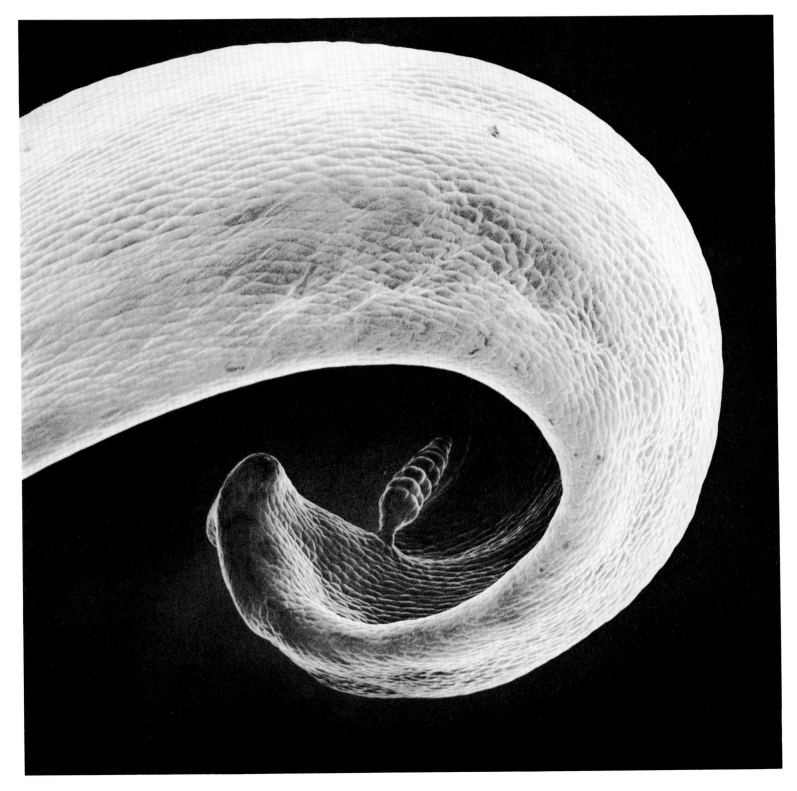

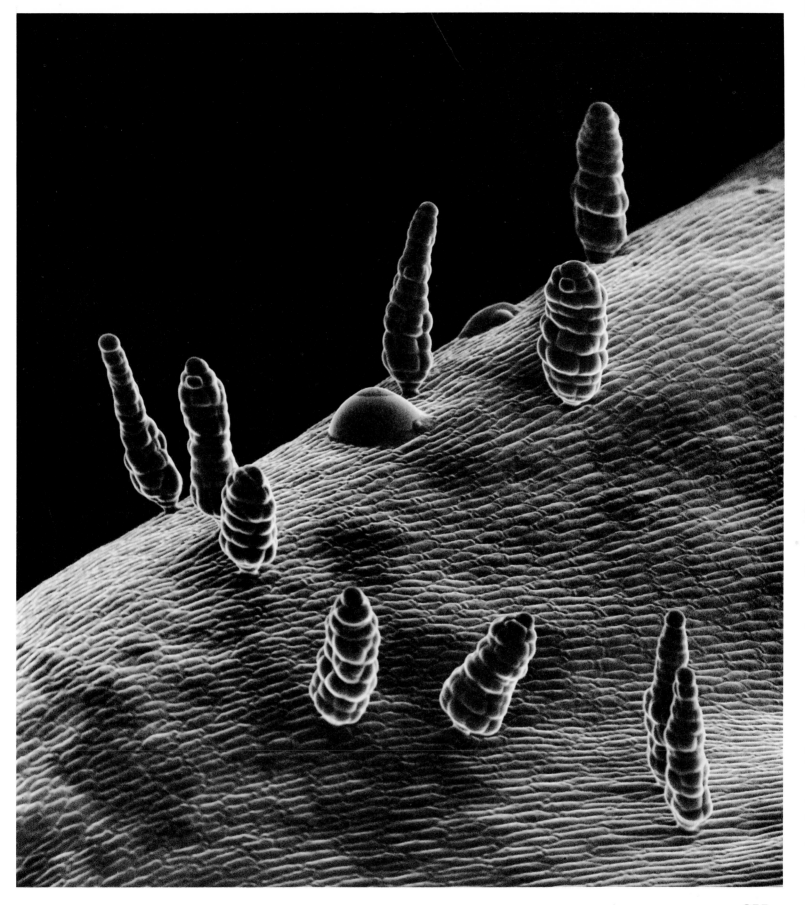

255 x

"Things on a Hibiscus Flower Pistil Stem"
Multicellular trichomes.

330 x

"Fertilization"
Bean flower (*Phaseolus vulgaris*—Family Leguminosae)
The pistil has been fertilized with pollen from the stamens.

74

"Untitled"
Helxine, Baby's tears (*Soleirolii*—Family Urticaceae)
Budding flower (center) and leaves.

26 x

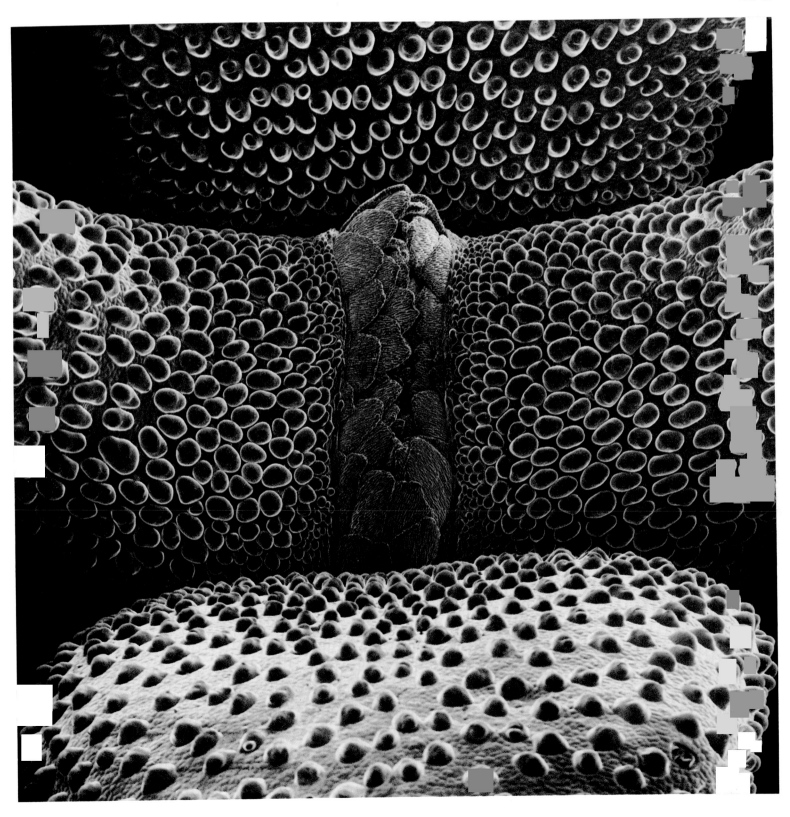

A California Wild Flower
Unidentified.

41 x

40 x

Wild Buckwheat Flower
(*Eriogonum Fasciculatum*—Family Polygonaceae)

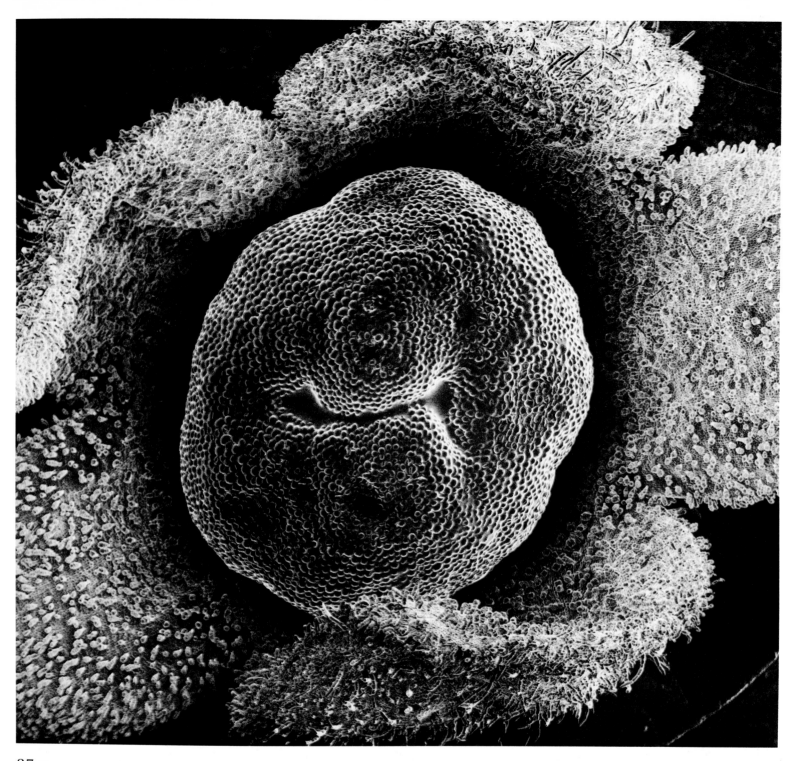

27 x

Wild Cucumber Flower
(*Marah Fabaceus*—Family Cucurbitaceae)

Scrub-Oak Leaf
(*Quercus Dumosa*—Family Fagaceae)
Underside of leaf showing many stellate (star-shaped) trichomes.

60 x

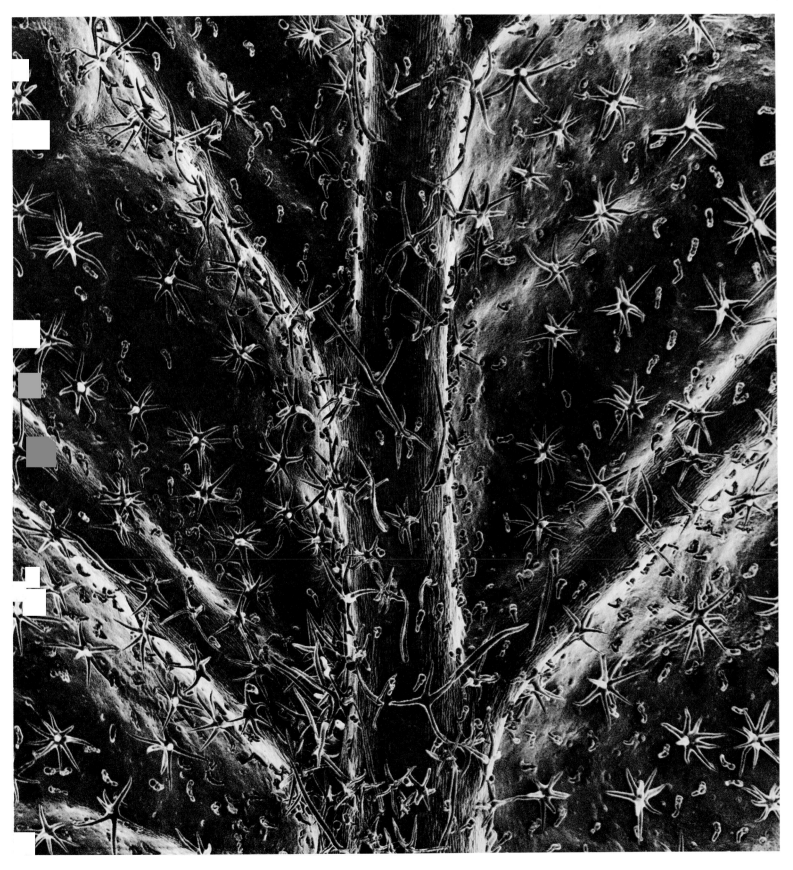

Coleus Leaf
(Family Labiatae)
Top surface of leaf.

66 x

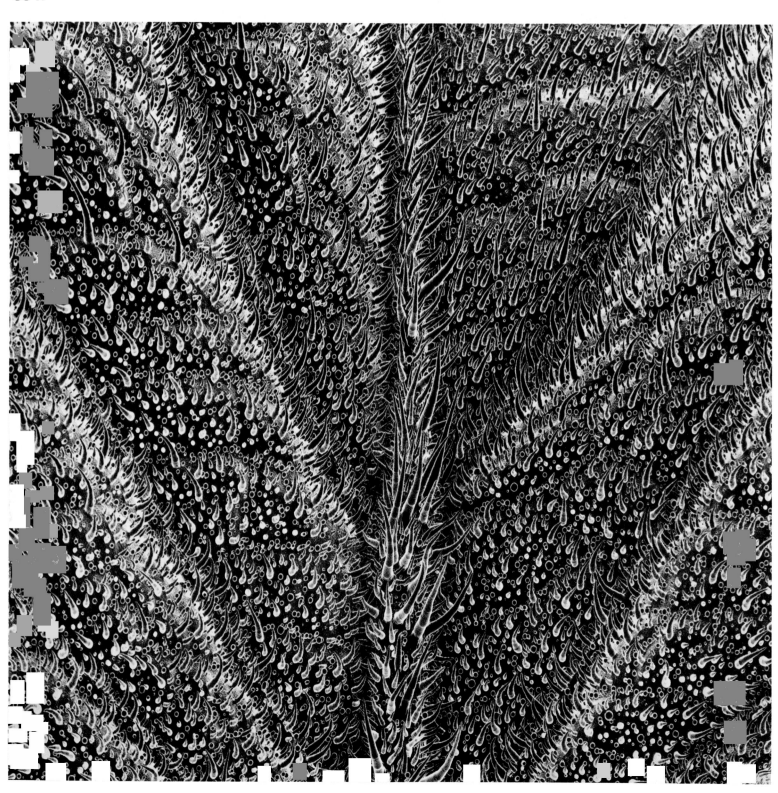

80

300 x

Coleus Leaf Closeup
(Family Labiatae)
Detail of picture at left. The surface is covered with glandular and non-glandular trichomes and hairs. The glandular trichomes are rounded and exude the plant resin.

180 x

Coleus Leaf Underside
(Family Labiatae)

Fennel Flowers
(*Foeniculum vulgare*—Family Unbelliferae)
These flowers are beginning to bloom.

60 x

Spotted Spurge

(*Euphorbia Supina*—Family Euphorbiaceae)
Flowers, leaves, and seed pods (fruits). A weed often found growing up through the cracks in sidewalks.

80 x

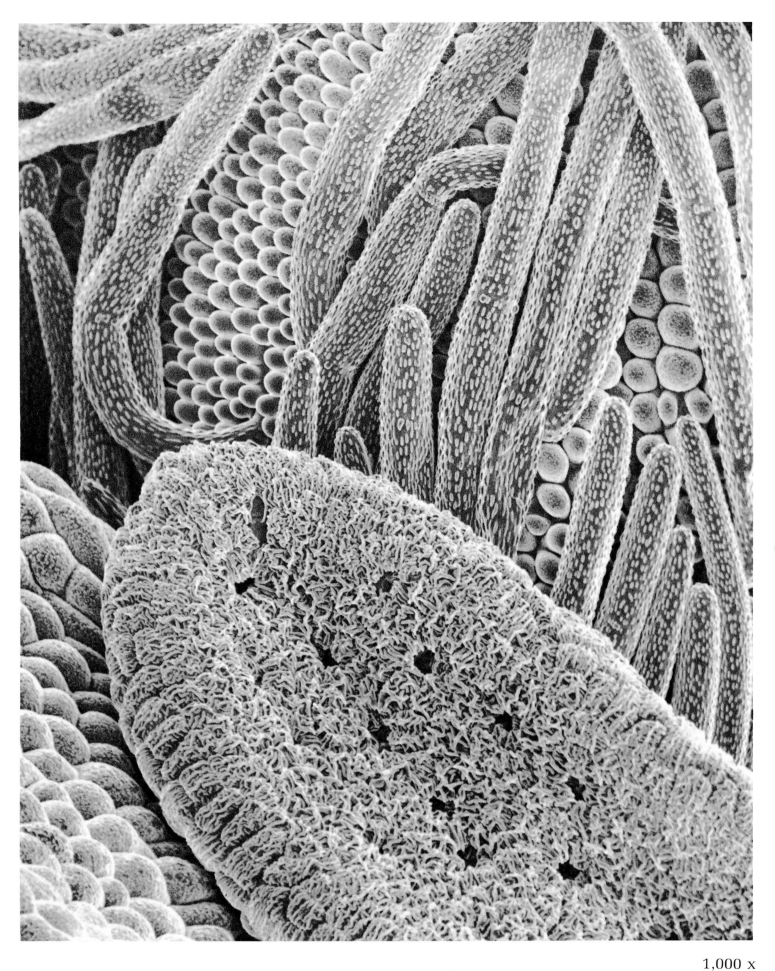

1,000 x

Spotted Spurge II
(*Euphorbia Supina*—Family Euphorbiaceae)
Hairs, cells, and parts of flower petal.

85

92 x

Daisy Flowers
African Daisy (*Arctotis*—Family Compositae)
Multiple unopened disc flowers of a Daisy's center. They open to release the pistil
and stamen inside. Composite flowers.

Daisy Flower (Prematurely Fertilized)
African Daisy (*Arctotis*—Family Compositae)
Note pollen in the center of unopened disc flower.

266 x

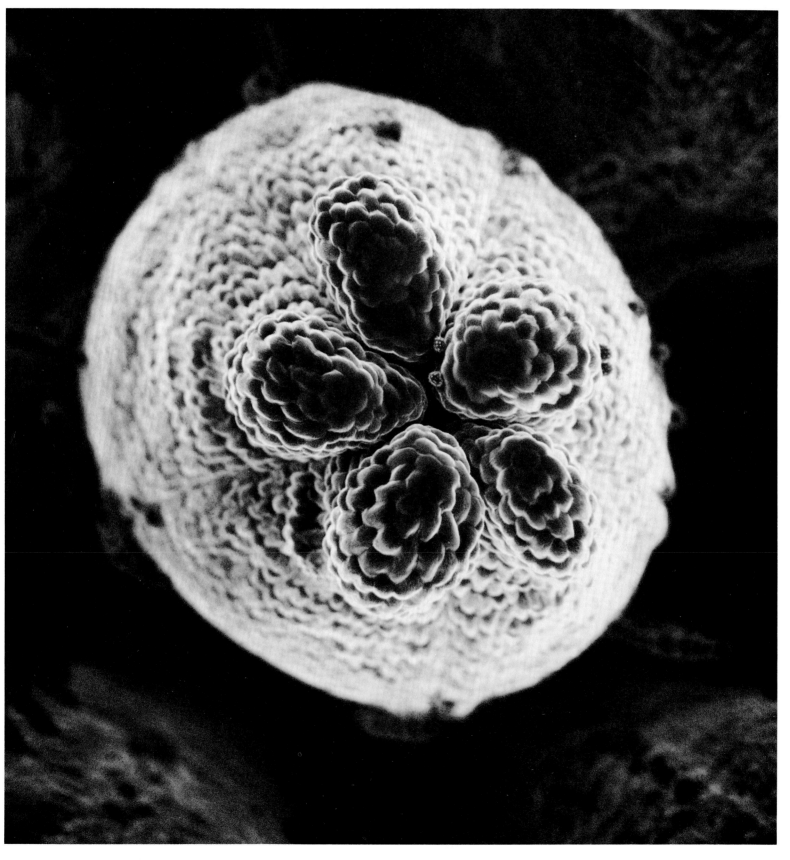

"Sea of Cells"
African Daisy (*Arctotis*—Family Compositae)
Flower petal surface.

175 x

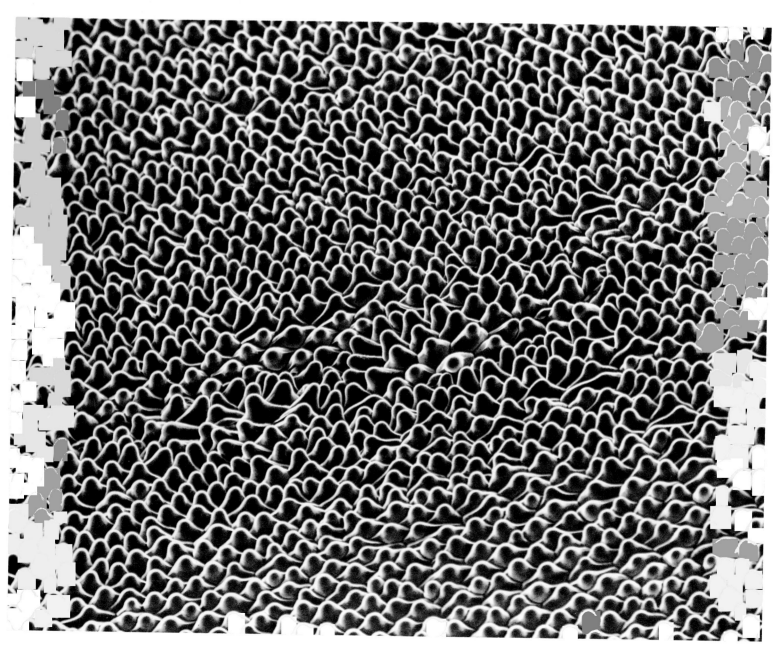

88

93 x

"Tarweed Star"
Yellow Tarweed Flower (*Hemizonia Virgata*—Family Compositae)
Showing the petals of several ray flowers—note glandular trichomes in center.

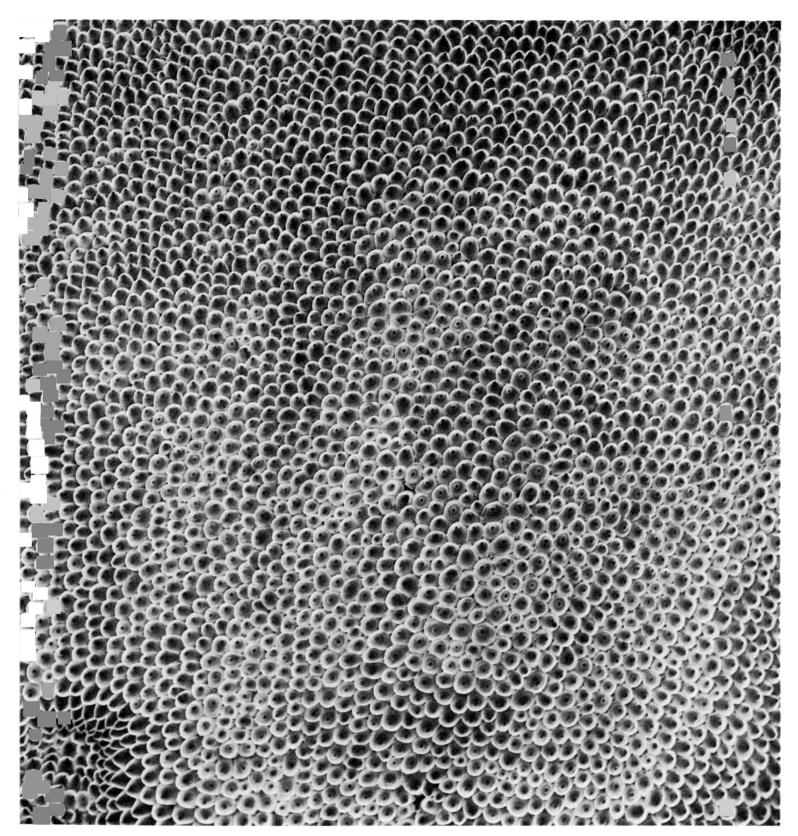

185 x

"Surface Study"
(*Gladiolus*—Family Iridaceae)
Gladiola flower petal.

Gladiola Flower Petal II

(*Gladiolus*—Family Iridaceae)
Note individual cells.

460 x

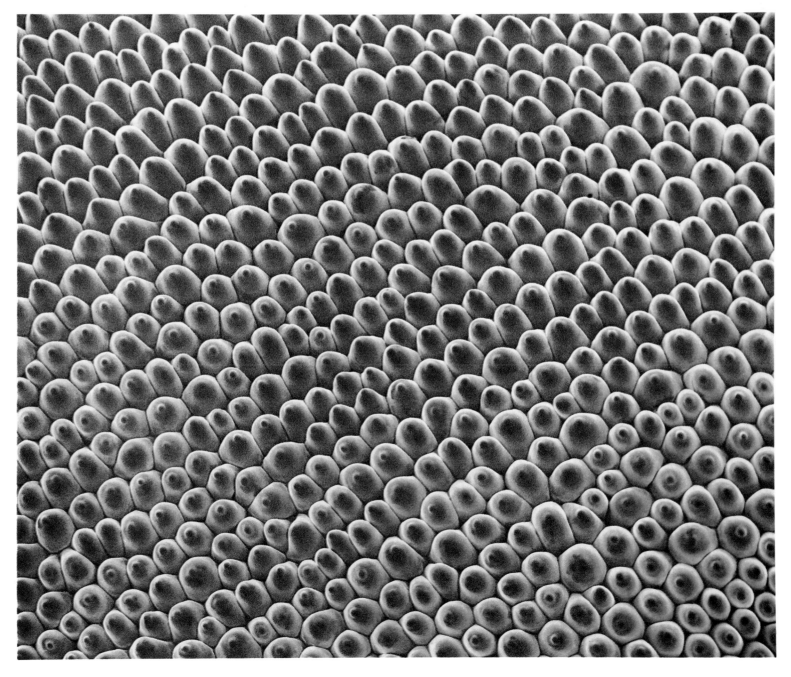

"Pittosporum"
Victoria Spring Tree (Family Pittosporaceae)
Flower stem . . . dotted with stomata, the small pores through which a plant
"breathes." A "guard cell" on each side of the pore serves to open and close it.

302 x

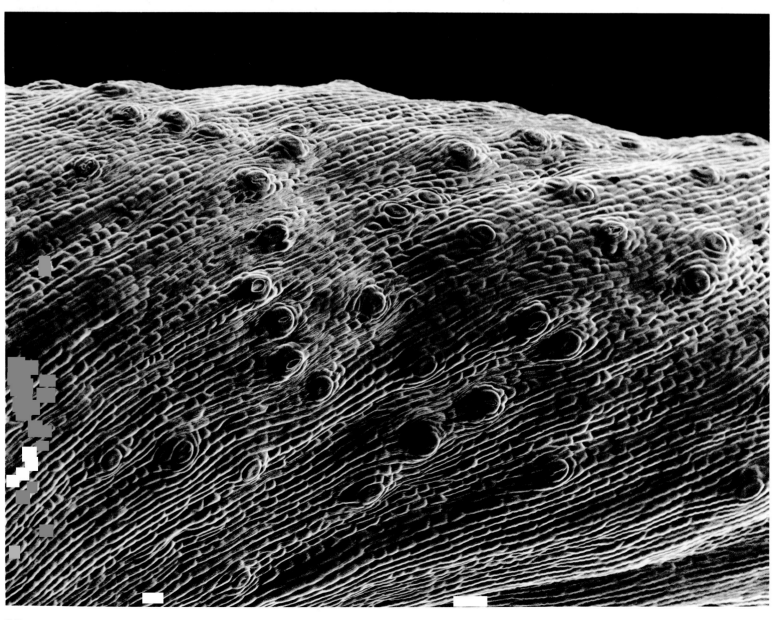

337 x

"Encounter"
Lantana (Family Verbenaceae)
Lantana flower and stem surface.

222 x

Marijuana Leaf I
(*Cannabis Sativa*—Family Moraceae)
A young leaf.

Marijuana Leaf II

(*Cannabis Sativa*—Family Moraceae)
Closeup of upper left corner of **Marijuana Leaf I**. Both photographs showing the cystolyth hairs (hairs containing calcium carbonate and other organic materials) and two kinds of resin nodules (they are: *capitate trichomes*—those having heads; and *peltate trichomes*—those which are round and close to the surface; both are glandular).

633 x

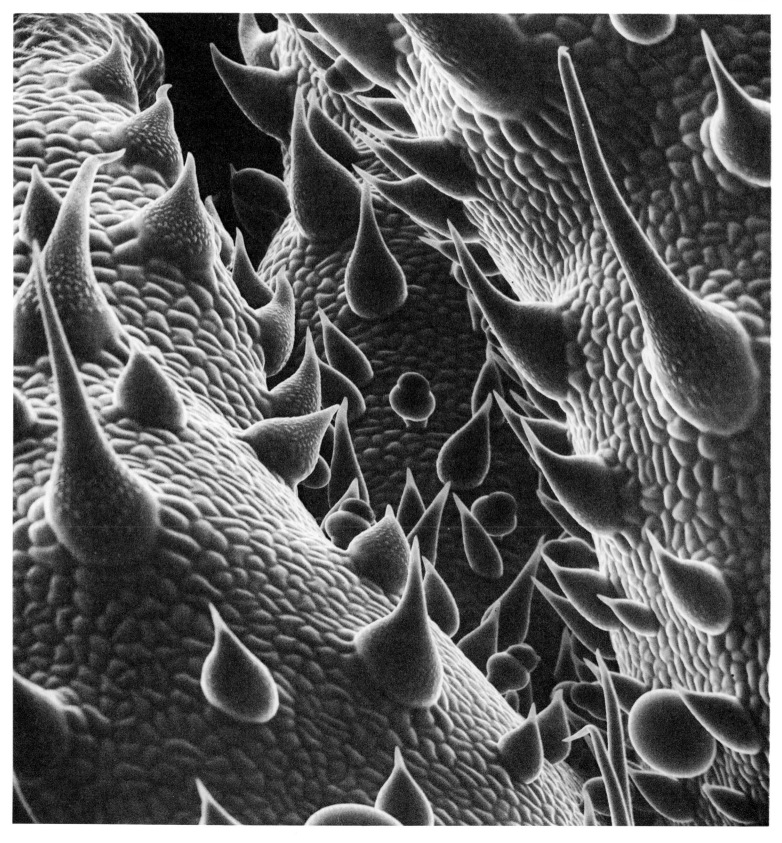

Marijuana Leaf III
(*Cannabis Sativa*—Family Moraceae)
Close view of a young leaf.

830 x

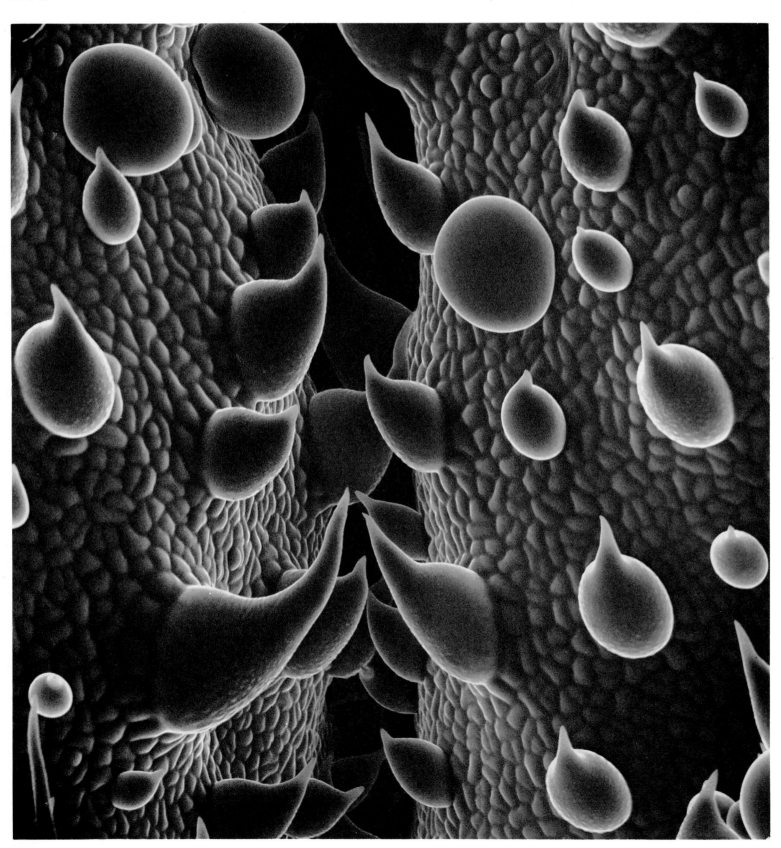

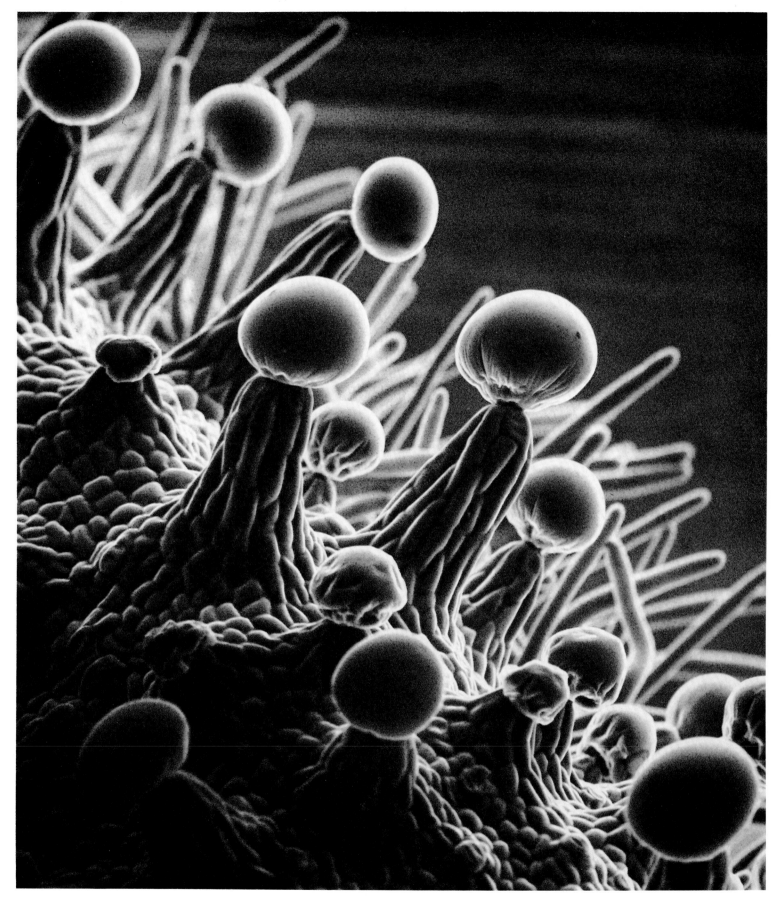

1,015 x

"High Power"
(*Cannabis Sativa*—Family Moraceae)
Resin nodules on a female Marijuana flower. These nodules are located on the bract which covers and protects the developing seed. They are glandular, capitate, multi-cellular trichomes that secrete the plant's concentrated drug.

STILL LIFE AND OTHER COMPOSITIONS

This section contains the photographs of both inanimate things and of life-forms whose imagery has been manipulated in a manner different from those of the first two sections. Little need be said about the inanimate subjects except that they too have a "life" of sorts—one could say that they describe some process or phenomenon in that they have a history and tell a story.

In some cases I find that I prefer a composition with its lights and darks reversed. The "mosquitoscape" is an interesting study of both reversal of its mirror image symmetry and of its image polarity.

50 x

"Landscape of a Famous Aspirin Tablet"
(gold coated)

1,250 x

Detail of Aspirin Tablet
From the center of the "Y". Note at least three different ingredients. (gold coated)

Silicon Carbide

Fractured surface of a silicon carbide crystal. This is one of the hardest materials known (ranking 4th); it withstands very high temperatures and is commonly used as an abrasive.

125 x

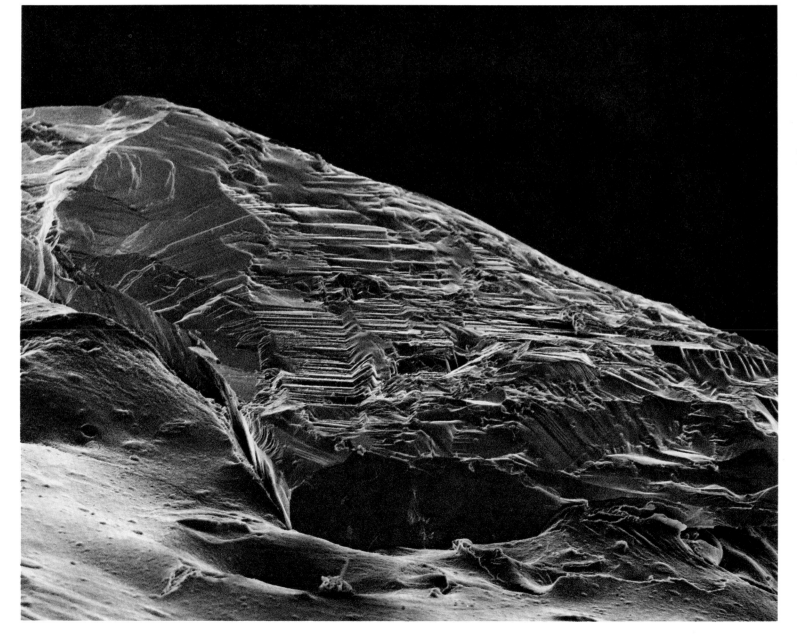

Salt Crystals
Table salt (gold coated)

135 x

242 x

"Salt Canyons"
Sea salt crystal. (gold coated)

96 x

Paper
Showing many fibers (gold coated)

Fabric
Fiberglass fabric. (gold coated)

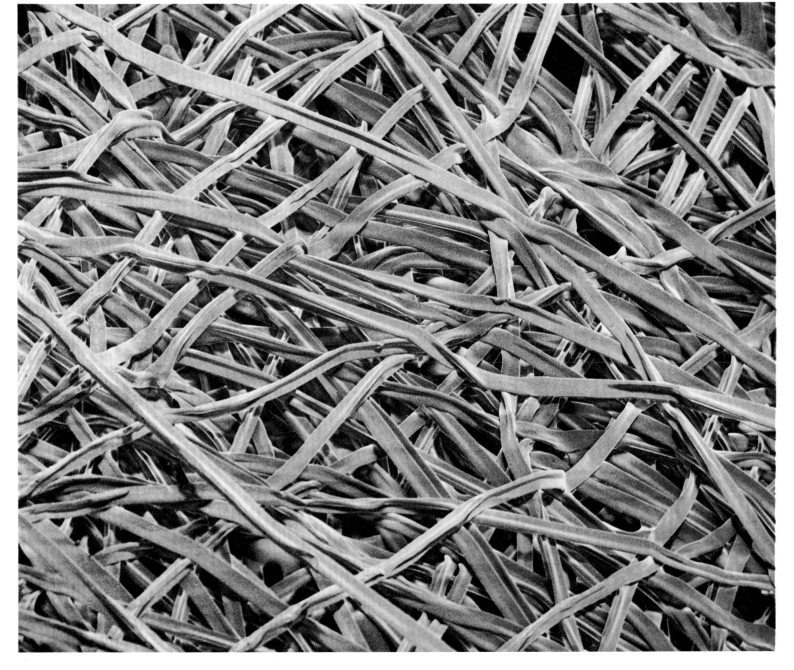

A Razor's Edge
Looking down (at an angle) at the edge of a razor blade.

1,000 x

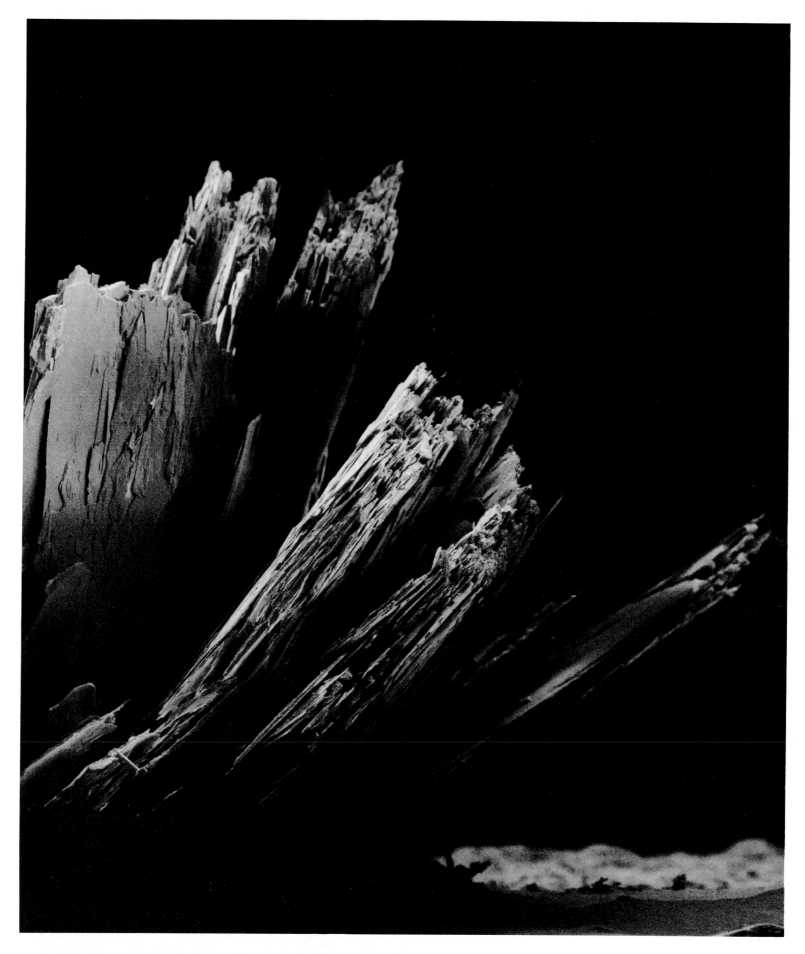

Cocaine
Commercial "flake" crystals. (gold coated)

2,880 x

"Y" Modulation I
Amplitude modulated image display (the usual picture is intensity modulated) of a Candytuft flower petal. Graphic display, electronically manipulated. Low scan density (few lines).

"Y" Modulation II

Same as previous photograph but with high scan density (many lines), varied contrast, and vertical expansion.

1,550 x

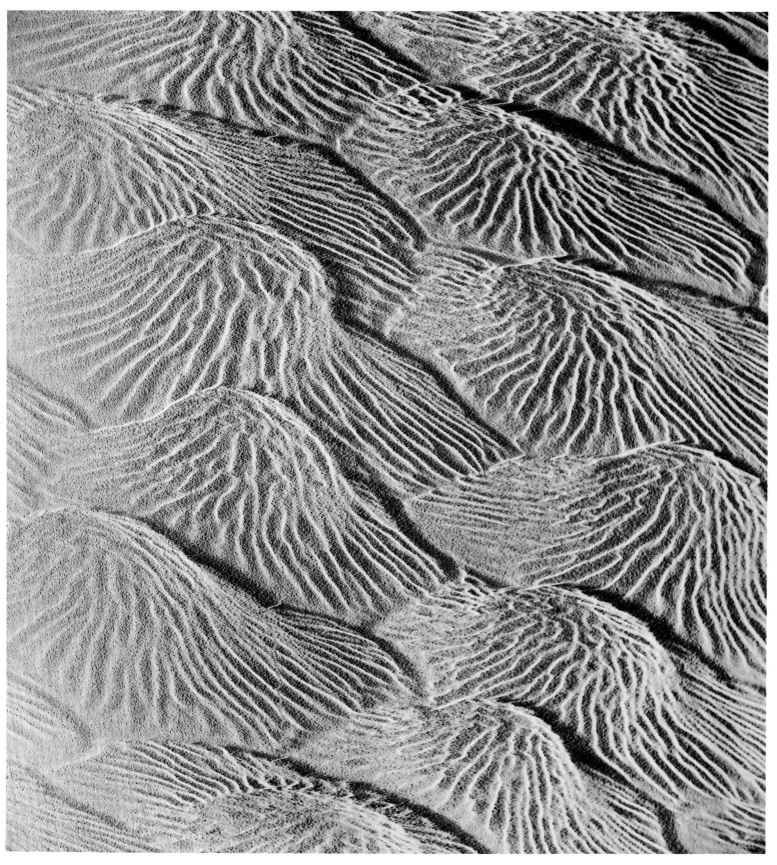

"Images in Talc"
Commercial baby powder. (gold coated)

1,255 x

8,300 x

"Spinout"

Corner of a highly polished glass-ceramic substrate with vacuum deposited layers of, first, copper; then nickel-iron alloy (note the dendrites of Ni-Fe overlapping the copper layer because of a different angle of deposition). Finally, on the top is a layer of photo-resist chemical which has been spun out to the edge in a centrifuge (note the ripples at the edge of the resist). Used for making special electronic devices. (gold coated)

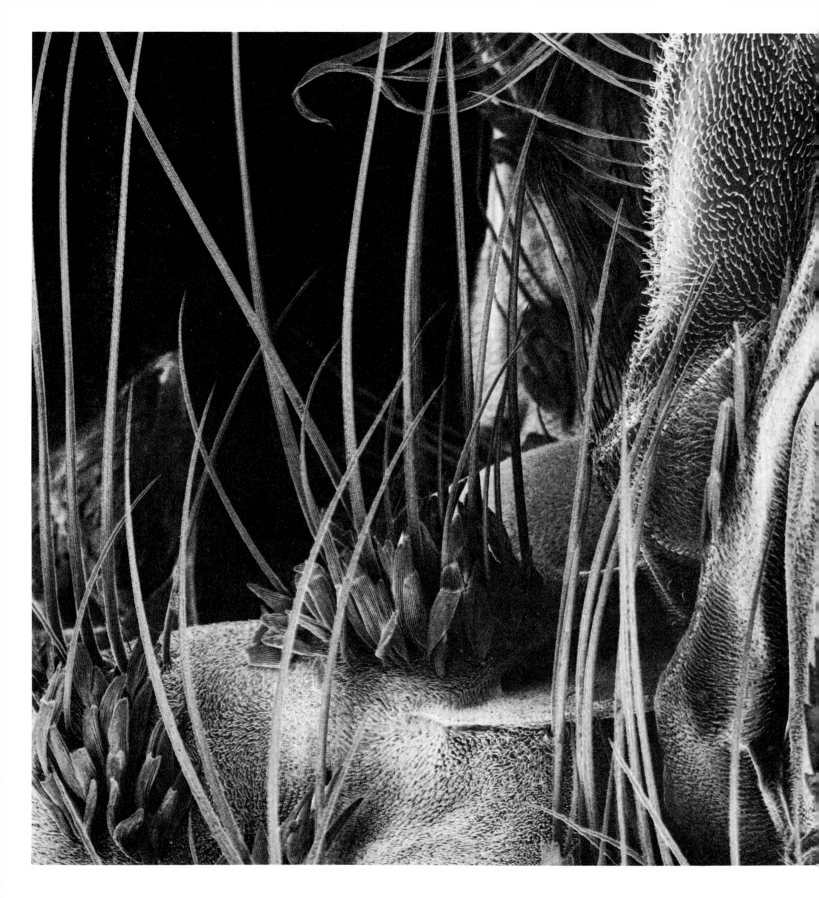

445 x

Symmetry Study I
A "Mosquitoscape" Showing hairs and scales on the thorax.

114

445 x

Symmetry Study II
A "Mosquitoscape" Showing hairs and scales on the thorax. (Reversed and negative image)

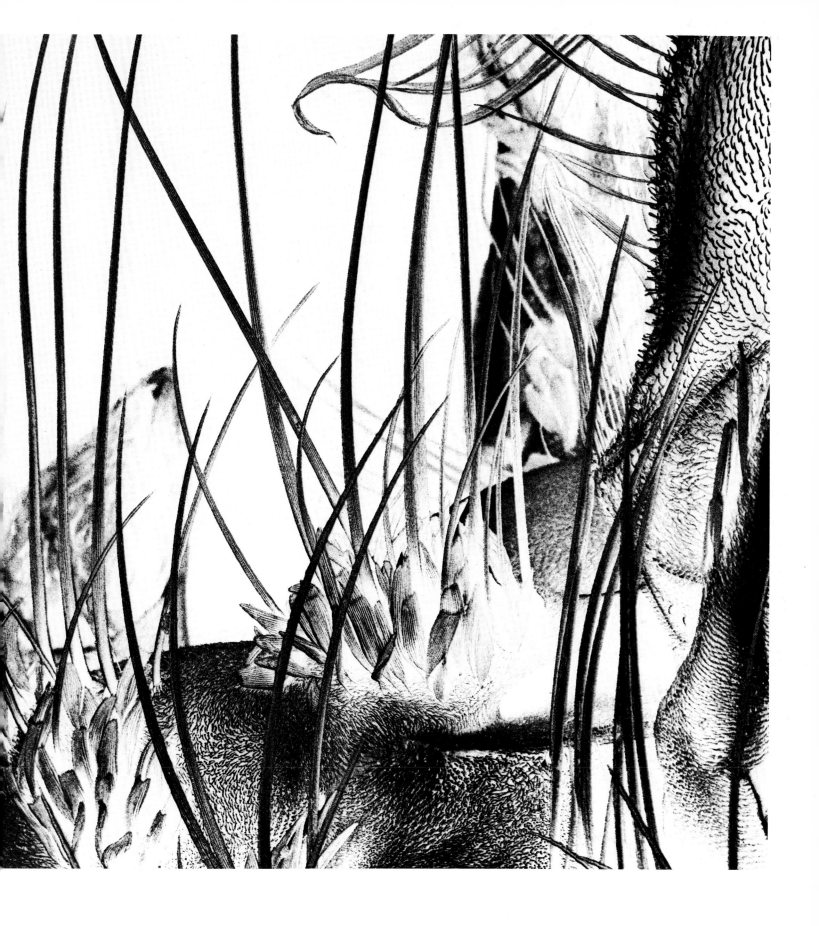

50 x

Sweet Alyssum Flowers
(*Lobularia*—Family Cruciferae) (negative image)

White Mulberry Tree Pistillate Flower
(*Morus Alba*—Family Moraceae) (negative image)

255 x

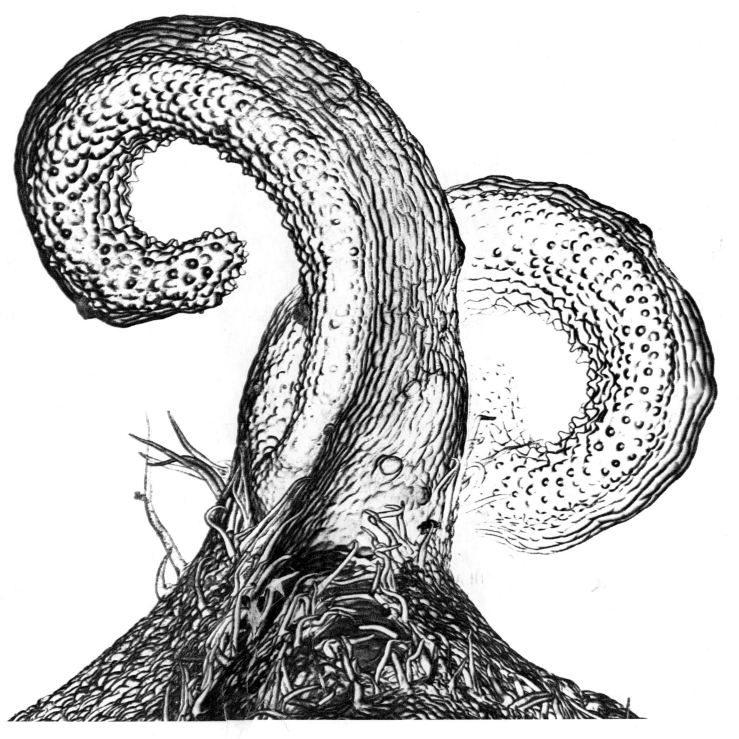